PRAISE FOR *NANOCONVERGENCE*

"*William Bainbridge is an original thinker who na... ences to emerging technologies and societal aspiratio... ... ..., ... the reader all that is essential in the historic and rapid change toward science and technology convergence.*"

—Mihail C. Roco, Ph.D., original chairman of the U.S. National Science and Technology Council's subcommittee on Nanoscale Science, Engineering and Technology, key architect of the National Nanotechnology Initiative, and currently senior advisor for Nanotechnology at the National Science Foundation

"*This book provides a sweeping, yet intimate, overview of an important, emerging area of science and technology—nanotechnology and its convergence with other areas of science and engineering. In Nanoconvergence we are provided with a view of these developments as seen through the lens of the world of William Sims Bainbridge, a visionary scientist and scholar, who has helped to frame and nurture nanoconvergence. His personal history and interests are endlessly fascinating, and include science fiction, space flight, religious cults, videogames, and a host of other areas and topics. His knowledge is extraordinary and includes expertise in the field of nanotechnology and related sciences, including biology, cognitive, behavioral and social science, and information technology. Further, he knows many of the players, including some who were mentors, others who are colleagues, and others whose funding he supervised. The strength of this book is the strength of Bainbridge's extensive, connected network, rooted in scientific, technological, and societal concerns.*

It is rare to find someone who brings to the table such breadth and depth of knowledge, spanning so many of the sciences, from physics through cognition. Bainbridge is a Renaissance man who is helping to both create and elucidate the potential future worlds that confront us. Ultimately, he is a visionary who is building a roadmap for a future that we can all help to shape. He is to be commended for sharing both this map and his journey with us."

—Philip Rubin, Ph.D., CEO, Haskins Laboratories

"*In a world of increasing specialization, Bainbridge offers a refreshing alternative perspective of the way nanoconvergence will help unify disparate areas of knowledge and fuel a next generation of innovation. The integration of historical and forward-looking insights, firmly grounded in the people and projects of the present, made this an enjoyable read. With Nanoconvergence, Bainbridge joins the ranks of the few authors who have succeeded in integrating insights from far flung fields of science and technology into a compelling human story.*"

—James C. Spohrer, Ph.D, Director, Services Research and Innovation Champion, IBM Almaden Research Center

Nanoconvergence

Nanoconvergence

The Unity of Nanoscience, Biotechnology, Information Technology, and Cognitive Science

William Sims Bainbridge

PRENTICE
HALL

Upper Saddle River, NJ • Boston • Indianapolis • San Francisco
New York • Toronto • Montreal • London • Munich • Paris • Madrid
Capetown • Sydney • Tokyo • Singapore • Mexico City

Many of the designations used by manufacturers and sellers to distinguish their products are claimed as trademarks. Where those designations appear in this book, and the publisher was aware of a trademark claim, the designations have been printed with initial capital letters or in all capitals.

The author and publisher have taken care in the preparation of this book, but make no expressed or implied warranty of any kind and assume no responsibility for errors or omissions. No liability is assumed for incidental or consequential damages in connection with or arising out of the use of the information or programs contained herein.

The publisher offers excellent discounts on this book when ordered in quantity for bulk purchases or special sales, which may include electronic versions and/or custom covers and content particular to your business, training goals, marketing focus, and branding interests.

For more information, please contact:

U.S. Corporate and Government Sales
(800) 382-3419
corpsales@pearsontechgroup.com

For sales outside the United States, please contact:

International Sales
international@pearsoned.com

This Book Is Safari Enabled

The Safari® Enabled icon on the cover of your favorite technology book means the book is available through Safari Bookshelf. When you buy this book, you get free access to the online edition for 45 days.

Safari Bookshelf is an electronic reference library that lets you easily search thousands of technical books, find code samples, download chapters, and access technical information whenever and wherever you need it.

To gain 45-day Safari Enabled access to this book:

• Go to http://www.prenhallprofessional.com/safarienabled
• Complete the brief registration form
• Enter the coupon code NNDL-DBXD-VRWP-3FE3-CJ89

If you have difficulty registering on Safari Bookshelf or accessing the online edition, please e-mail customer-service@safaribooksonline.com.

Visit us on the Web: www.prenhallprofessional.com

Library of Congress Cataloging-in-Publication Data

Bainbridge, William Sims.
  Nanoconvergence : the unity of nanoscience, biotechnology, information technology, and cognitive science / William Sims Bainbridge.
      p. cm.
  Includes bibliographical references and index.
  ISBN 978-0-13-244643-3 (pbk. : alk. paper)
  1. Nanotechnology. 2. Technological innovations. I. Title.
  T174.7.B32 2007
  620'.5—dc22

                                                                        2007014262

ISBN 13:   978-0-13-244643-3
ISBN 10:   0-13-244643-X
Text printed in the United States on recycled paper at RR Donnelley in Crawfordsville, Indiana.
First printing, June 2007

*To Julia Constance Campos Bainbridge, in the hope
that she will experience all the good possibilities
imagined here, and none of the bad ones.*

Contents

Preface

This book explores the future of science and technology, and their implications for human beings. It is based on the insights of hundreds of scientists and engineers working at the cutting edge of research, as seen through the eyes of a social scientist who worked alongside them to organize, write, and edit a series of influential government-sponsored and independent reports. Although I have made every effort to be balanced and comprehensive, this book is not a sterile exercise in abstraction and objectivity. Rather, it seeks to provide information that will be both fascinating and useful for students, entrepreneurs, investors, fellow scientists or engineers, and people in many walks of life who want to understand how their work and their world will change in coming decades.

One of the scariest questions for young people is this: "What will you be when you grow up?" Sometimes people nearing retirement age joke, "I still don't know what I'm going to be when I grow up!" Often *be* means *do,* and the question really refers to selecting a career and finding a job. More broadly, the question might refer to what kind of person you or I might become, in whatever span of life is left to us on this spinning planet. However the question is defined, it cannot be answered in isolation. A person cannot simply decide to become a blacksmith, elevator operator, or spaceship pilot. The economy and the technological culture must provide such jobs, or no one can get them. Contrary to predictions, the trade of blacksmithing did not completely disappear, although its role in society has been greatly diminished. I suppose elevator operators became security guards—and I wonder if they considered that change to be a demotion or a promotion. I don't know what happened to all of the prospective spaceship pilots. My point is that the nature of technical work, and the nature of the world in which we all live, will change radically in the future, because science and technology have entered an era of fundamental transformation.

At the time of the "dot-com crash" nearly a decade ago, computer professionals used to joke, "Now we'll find out how many computer programmers the world really needs." The implication was that data processing had been going through a technological revolution, but after the guns had fallen silent, there might not be much action anymore. Everyone in the field had noticed that big companies and government agencies had been producing their own electronic data systems, often at great cost and with dismal results. Soon, it was believed, they would admit that the desire to have their own proprietary systems was a dysfunctional status obsession and begin to buy their software off the shelf—just as everyone else was already doing. In the early 1980s, very small companies could succeed while writing software for the consumer market, but since then a shakeout had occurred in small business and home office software. By way of analogy, in the beginning of the twentieth century, scores of small companies set out to make automobiles, but within half a century the overwhelming majority had ceased to exist. Perhaps by 2010, every business on the face of the Earth could make do with Microsoft Office.

This issue raises two questions very germane to the topic of this book: "What is computer science?" and "How can it continue to progress?" Computer science is not simply programming, nor is it the more exalted profession of software engineering, although both entities depend on it. Nor is computer science merely a branch of electrical engineering, although many people who call themselves computer scientists have a degree in "EE." Rather, computer science is an incomplete convergence of mutually supportive fields that cooperate to produce the hardware, software, and management systems required to process information, including in consumer areas such as the World Wide Web and online games, as well as in service of corporations and government agencies. As "comp-sci" matures, it draws more and more fields into it. Early on, it attracted many mathematicians; today, it needs the expertise of members of the cognitive science field and the social sciences. As this unification progresses, the field should probably be renamed simply *information science*. Indeed, this term is already in wide circulation, where it is used to encompass all forms of communication, whether or not they are supported by electronic devices.

We cannot be sure how much longer the electronic hardware will continue to progress. In the past, hardware advances both permitted and demanded software advances, and the evolution of the two together enabled entirely new applications. When I entered Yale University as a physics major in 1958, it was widely believed that two prominent application areas, nuclear technology and space rocketry, would rise still further to transform the world. This proved to be a miscalculation: Within 15 years, both areas had largely stalled. We still need nuclear and aerospace engineers, but now they work primarily as the paid minions of corporate executives, with very limited scope for personal innovation. The same is true for most computer professionals in large organizations. Even so, the information area has kept lively because individual entrepreneurs and small companies have continued to develop new approaches and applications. "A revolution every five minutes" is a slight exaggeration, but this period of growth and discovery could end at any time.

So what is a person to do? What I did, when I was young, was stumble from field to field for a few years, before realizing that as a social scientist I could keep innovating by applying my growing professional experience to a series of different topics, each appropriate for the decade in which I was working on it. At a recent computer science convention, a couple of corporate recruiters told me they were looking for students who knew exactly what specialty they wanted to work in, and who were gaining the precise expertise required for that niche. I was horrified to hear this. What will these companies do with these people when their specialties are no longer needed in a few years? Fire them, probably.

A young person seeking a career in science or engineering today should start from the hopeful premise that the fundamental things he or she is really interested in will remain important decades later. But such a person cannot assume that particular narrow technical fields or job classifications will still exist even one decade in the future. The fact that many of the best opportunities will exist at the boundaries of fields does not mean that a student should avoid exploring one field deeply. For many, a "T-shaped" expertise will be best—that is, deep in one area but also covering adjacent areas. Often, a corporation or other technical organization will value highly a person who has solid expertise in a field central to its work, but who also possesses enough expertise in adjacent areas to contribute to a multidisciplinary team, or even to promote transfer of new ideas from one field to another. Opportunities for such a person become especially great when an entire new field is opening up.

Many physicists who happened to be mathematically inclined became computer scientists simply by redefining the expertise they already had. Others, who were better with chemistry than math, became materials scientists, and more recently redefined their expertise as nanoscience. As this book will demonstrate, nanotechnology is converging with biotechnology and information technology. Great opportunities exist for people who are prepared to build the bridges between those fields today.

Does this transdisciplinary philosophy place unreasonable demands on students, asking them to add extra work to the full-time job of learning one field well? Not necessarily, if their teachers also evolve with the changing conditions in science and technology. Much of the "expertise" in many fields consists of brute, dumb facts, often in the form of unnecessary nomenclatures. The unification of the sciences and branches of engineering requires a transformation of their styles and cultures. Part of that transition will be achieved by easy-to-use information technology systems that replace the arcane technical handbooks of the past. Part of it will be achieved by new terminology and analytic or design procedures that can be applied broadly across fields. And part of it will be achieved by the development of new professions specifically designed to bridge between specialized branches of expertise.

When I earned my doctorate in sociology from Harvard University in 1975, with a dissertation on the social history of the space program, I was lucky to get a job in the tenth-ranked sociology department in the country, because the job market was in the process of crashing. Enthusiasm for the social sciences began to dwindle at that time, and today the social sciences (except economics, if you want to count that "rich" field among the social sciences) have less influence than they did in the 1950s and 1960s. Coincidentally, 1975 also marked the end of the remarkably vigorous post-war growth of U.S. universities. Put bluntly, it is hard to name any clear-cut discoveries achieved in the social sciences comparable to the feats achieved in genetics, for example, over the same period. And yet, public confidence in political leaders is justifiably low at the present time, and advanced societies face many policy decisions, including some concerning which technologies to promote or prohibit. We would be better off today if the social sciences were more influential, and if they had earned that position on the basis of solid achievements based on actual scientific discovery. Ultimately, winning such respect will require the social sciences to become integrated with the cognitive sciences, on the basis of a shared understanding of human behavior.

This book has two themes. One is clearly stated in the title: *Nanoconvergence.* Today, nanotechnology is converging on the one side with information technology, and on the other side with biotechnology. The convergence of information technology with biotechnology is making it possible to build new technologies on the basis of cognitive science, all enabled by nanotechnology. The second theme is perhaps less clearly stated in the identity of the author of the book, a social scientist who became an information or computer scientist and worked with the National Nanotechnology Initiative. Technological convergence requires a social awareness if it is to benefit people, and that awareness can best be achieved by reviving social science on the basis of its convergence with the other fields.

While useful for students who face career choices, this book is not narrowly aimed at people who are deciding what they want to be when they grow up, except in the sense that we all must negotiate shifts in our identity in this changing world, hoping we all grow intellectually so long as we live. Whether as investors, managers, consumers, or citizens, we will all face choices related to science and technology. This book is intended to be a resource for people who are contemplating many kinds of choices, and for people who are interested in understanding the world around them. It seeks to put the reader into communication with hundreds of scientists and engineers, and with the hundreds of social scientists and philosophers who have collaborated with them, so as to share their excitement and wisdom about the coming convergence.

About the Author

Williams Sims Bainbridge is the author of 17 scientific books, about 200 shorter works, and a number of software programs in the social science of technology and other fields. After earning his doctorate from Harvard University with a dissertation about the space program, he taught sociology at the University of Washington and other universities for twenty years before joining the National Science Foundation to run its sociology program. In 2000, he moved over to NSF's Directorate for Computer and Information Science and Engineering, where he currently is a grant program officer in the Human-Centered Computing area. For years he represented the social sciences on the major interagency high-technology funding initiatives, from High Performance Computing and Communications to the National Nanotechnology Initiative. He has played leading roles in the two most important efforts to anticipate the societal implications of nanotechnology, and in four related projects charting the potential for transformation of science and engineering by the unification of the NBIC fields: nanotechnology, biotechnology, information technology, and new technologies based on cognitive science.

Chapter 1

Convergence at the Nanoscale

Nanoscience and nanotechnology are not merely "the next big things," offering investors the chance to get in on the ground floor of new industries. More importantly, they promote the unification of most branches of science and technology, based on the unity of nature at the nanoscale. Already, information technology incorporates hardware with nanoscale components, and biotechnology is merging with nanotechnology in many areas. Indeed, unless these technologies converge, further progress in most fields will be impossible. More controversially—but also more significantly—the convergence is prepared to encompass cognitive science. This vast unification is often called NBIC, from the initials of its four main components: Nanotechnology, Biotechnology, Information technology, and Cognitive science. The result will be new cognitive technologies that promise to put the behavioral and social sciences for the first time on a rigorous foundation.

This book reports the latest developments in, and tantalizing possibilities related to, convergence at the nanoscale. The perspective taken here is that of a social and information scientist who has been centrally involved in major collaborative projects to assess the implications of nanoscience and nanotechnology.

THE MEANING OF "NANO"

Convergence of NBIC technologies will be based on *material unity at the nanoscale* and on *technology integration from that scale*. The building blocks of matter that are fundamental to all sciences originate at the nanoscale—that is, the scale at which complex inorganic materials take on the characteristic mechanical, electrical, and chemical properties they exhibit at larger scales. The nanoscale is where the fundamental structures of life arise inside biological cells, including the DNA molecule itself. Soon, the elementary electronic components that are the basis of information technology will be constructed at the nanoscale. Understanding the function of the human brain requires research on nanoscale phenomena at receptor sites on neurons, and much brain research will be facilitated by nanoscale components in microsensor arrays and comparable scientific tools. Thus nanotechnology will play an essential role both in achieving progress in each of the four fields and in unifying them all.

Although perhaps everyone understands that "nano" concerns the very small, it is nevertheless difficult to get a picture of how small a nanometer really is: one billionth (thousand-millionth) of a meter. A billionth of a meter is the same as a millionth of a millimeter, and the smallest U.S. coin, the "thin" dime, is about a millimeter in thickness. If you were somehow able to shrink yourself down until you were only a nanometer tall, then in comparison a dime would seem to be 175 kilometers (about 100 miles) thick. The DNA in the cells of our bodies is between 2 and 3 nanometers thick, though as much as several millimeters long, so it has the proportions of a long piece of fine thread, curled up inside the chromosomes. Atoms and water molecules are smaller than a nanometer, whereas the wavelength of visible light ranges from approximately 400 nanometers at the violet end of the spectrum to approximately 700 nanometers at the red.

In 1960, the General Conference on Weights and Measures refined the metric system of measurement, among other things defining the nanometer as one billionth of a meter. Another unit for measuring tiny distances was already widely used in spectroscopy and nuclear physics, the ångström, which is 0.1 nanometer. In principle, the ångström became obsolete in 1960, but, in fact, it is still used today.

During the next few years, "nano" concepts became widely disseminated throughout the cultures of civilized nations. For example, the 1966–1967 sci-fi television series *Time Tunnel* used the word "nanosecond" as part of the countdown to operate its time machine: "One second, millisecond, microsecond, nanosecond!" The term "nanotechnology" was apparently first used by Professor Norio Taniguchi of Tokyo Science University in a 1974 paper, in which it described the ultimate standard for precision engineering."[1]

Recently, people have been coining "nanowords" at a furious pace. For an article I published in *The Journal of Nanoparticle Research* in 2004, I counted titles containing "nano" on the Amazon.com website, finding 180 books that fit the bill.[2] Some contain two "nano" words, such as *Societal Implications of Nanoscience and Nanotechnology*, edited by Mihail ("Mike") C. Roco and myself in 2000. Altogether, there were 221 "nano" words in the titles of these 180 books. Nanotechnology is most common, appearing 94 times. Nanostructure (or a variant like nanostructured) appeared 28 times, and nano, 18 times. These words appeared five times: nanocomposite, nanofabrication, nanomaterials, nanophase, nanotribology, nanoscale, and nanoscience.

Nanosystems appeared four times, and nanoengineering, nanoindentation, nanomeeting, and nanoparticles appeared three times each. Bionanotechnology appeared twice, as did nanocrystalline, nanoelectronics, nanometer, nanophotonics, nanotech, and nanoworld. Sixteen other words appeared once in the titles: nanobelts, nanobiology, nanocosm, nanodevices, nanoelectromechanics, nanolithography, nanomechanics, nanomedicine, nanometric, nanometrology, nanoporous, nanopositioning, nanoscopy, nanosources, nanotubes, and nanowires. Clearly, we are facing a nanocraze, nanofad, or nanohype.

This welter of words may bring pure nano into disrepute, because it seems to be claiming too much scope for the field. Nevertheless, it would be a mistake to think that nanotechnology is a specific technical approach, such as the fabled nanoscale robots that some visionaries imagine. Rather, the nanoscale is the region where many technologies meet, combine, and creatively generate a world of possibilities. The official website of the National Nanotechnology Initiative (NNI; www.nano.gov) defines the field as follows:

> Nanotechnology is the understanding and control of matter at dimensions of roughly 1 to 100 nanometers, where unique phenomena enable novel applications. Encompassing nanoscale science, engineering, and technology, nanotechnology involves imaging, measuring, modeling, and manipulating matter at this length scale. At the nanoscale, the physical, chemical, and biological properties of materials differ in fundamental and valuable ways from the properties of individual atoms and molecules or bulk matter. Nanotechnology R&D is directed toward understanding and creating improved materials, devices, and systems that exploit these new properties.

For NNI leader Mike Roco (Figure 1–1), the scientific challenges of this length scale are as immense as the technical opportunities:

> We know most about single atoms and molecules at one end, and on bulk behavior of materials and systems at the other end. We know less about the intermediate length scale—the nanoscale, which is the natural threshold where all living systems and manmade systems work. This is the scale where the first level of organization of molecules and atoms in nanocrystals, nanotubes, nanobiomotors, etc., is established. Here, the basic properties and functions of material structures and systems are defined, and even more importantly can be changed as a function of organization of matter via "weak" molecular interactions.[3]

Figure 1–1 Mihail C. Roco, Senior Advisor for Nanotechnology to the Directorate for Engineering, National Science Foundation. Mike has not only been the most forceful advocate for nanoscience and technology, but also originated many of the key ideas in NBIC convergence on the basis of considering the societal implications of nanotechnology.

When I first became professionally involved with nanotechnology, I was a member of the scientific staff of the Directorate for Social, Behavioral, and Economic Sciences of the National Science Foundation (NSF). Since 1993, I had been representing the directorate on computer-oriented cross-cutting initiatives, such as High-Performance Computing and Communications, the Digital Library Initiative, and Information Technology Research. I was a life-long technology enthusiast, having written three books about the space program, experimented with musical technologies from harpsichords to electronic tone generators, and programmed a good deal of educational and research software. Thus, when Mike Roco approached the directorate in 1999, seeking someone to represent the social sciences on the nanotechnology initiative he was organizing, I was excited to volunteer.

NANOTECHNOLOGY AND SCIENTIFIC PROGRESS

Unwittingly, I first encountered nanotechnology when I was a very small child. When I was four years old, I had the opportunity to visit the laboratory of multimillionaire and nuclear scientist Alfred Lee Loomis in his mansion at Tuxedo Park, New York. He showed me secrets that were too highly classified for an adult who might understand their importance. Loomis was a financier with some connection to my maternal grandfather's Wall Street law firm, but he was also a practicing physicist who played major roles in two high-tech programs that helped the Allies win World War II: the Manhattan Project, which developed the atomic bomb, and the Radiation Laboratory at Massachusetts Institute of Technology (MIT), which developed radar.[4] In his lab, Loomis first showed me a cup and then poured water into it; I was astonished to see that the water poured out again magically through the solid ceramic material. Only decades later did I realize that I had seen the fundamental secret of gas diffusion uranium isotope separation. It was my first introduction to nanotechnology.

There are several ways to obtain the fissionable material necessary to make an atom bomb. One of the first means developed relied on the separation of U235, the isotope of uranium suited for a bomb, from the unsuitable but much more common U238, using gas diffusion. Because they are isotopes of the same chemical element, the two cannot be separated by means of any chemical reaction. Instead, their slightly different physical properties need to be exploited to carry out the separation.

In this technique, uranium composed of both isotopes is chemically combined with fluorine to make uranium hexafluoride, which when heated becomes a gas. This gas is extremely corrosive and must be handled very carefully because of both its chemical properties and its radioactivity. For example, when uranium hexafluoride meets water, it generates hydrofluoric acid, which is so corrosive it can eat through glass.

The uranium hexafluoride is then passed through a porous barrier—a sheet of something with holes to allow the gas through—that slows the U238 down slightly, because it is slightly heavier. Although the exact details remain classified, the ideal average size of the pores is about 10 nanometers.[5] This is not just a matter of having holes that are exactly the right size to let U235 through yet block U238. A uranium atom is slightly less than one nanometer in diameter, and clustering six fluorine atoms around it does not produce a big molecule. The efficiency of the separation process is low, so it is necessary to cascade a large number of separation steps to enrich the uranium sufficiently for use in a bomb, and other methods are used today.

When Loomis showed me his sample of the gas diffusion barrier, in the form of a cup that could not hold its water, the first atomic bomb had not been detonated yet, and the word "nanotechnology" had not been coined. Nevertheless, even a child could see that his laboratory held secrets of the utmost importance. A sense of how far nano has come since those bygone days can be gained from the speeches given by six scientists when they accepted the Nobel Prize for great advances that enabled rapid development in nanoscience. The NNI website notes, "Nanoscale science was enabled by advances in microscopy, most notably the electron, scanning tunneling, and atomic force microscopes, among others. The 1986 Nobel Prize for Physics honored three of the inventors of the electron and scanning tunnel microscopes: Ernst Ruska, Gerd Binnig, and Heinrich Rohrer."[6]

The first electron microscope, which was built in 1931 by Ruska and Max Knoll, was hardly more powerful than a student's optical microscope, magnifying objects 400 times their diameter. Optical microscopes, however, remain limited by the rather long wavelengths of visible light (400–700 nanometers). In contrast, over a period of years, the resolving power of electron microscopes gradually sharpened until it reached deep into the nanoscale. The research by Ruska and Knoll was initially intended to refine oscilloscopes—devices used to measure fluctuating electric currents and signals, which were based on the same kind of cathode ray tube used as the picture tube in television sets before the introduction of flat screens. A cathode ray tube draws a picture on a fluorescent screen by scanning an electron beam over it. In 1929, Ruska became the first person to carry out experiments in which a well-focused electron beam actually cast images of a physical object in the beam's path. Two years later he developed an arrangement of focusing coils that permitted enlargement of the image—that is, the first electron microscope.[7]

Gerd Binnig and Heinrich Rohrer did not set out to develop a new kind of microscope, but rather sought to perform spectroscopic analysis of areas as small as 10 nanometers square. Interested in the quantum effect called tunneling, they were aware that other scientists were studying this phenomenon in connection with spectroscopy, and they began to think about how they might apply it in their own work. Binnig and Rohrer considered studying a material by passing a small probe with a very tiny tip over the surface so that electrons would tunnel across the gap. As they noted in their lecture accepting the Nobel Prize: "We became very excited about this experimental challenge and the opening up of new possibilities. Astonishingly, it took us a couple of weeks to realize that not only would we have a local spectroscopic probe, but that scanning would deliver spectroscopic and even topographic images, i.e., a new type of microscope."[8]

New measurement instruments and research methodologies are fundamental to the development of new fields of science and engineering. Once methods of research exist, then discoveries naturally follow. In 1985, Robert F. Curl, Jr., Sir Harold W. Kroto, and Richard E. Smalley discovered that carbon atoms can assemble into ball-shaped structures rather like the geodesic domes designed by architect Buckminster Fuller in the 1960s.[9] In recognition of the similarity, these assemblies of carbon atoms came to be called "buckyballs" or, more formally, buckminsterfullerenes (usually shortened to fullerenes). The

Figure 1–2 Superscale model of a fullerene, built by Troy McLuhan, on display in a virtual world. This nanoscale structure appears twice the height of a human being in the Science Center in the online environment called Second Life (http://www.secondlife.com/), illustrating the many convergences between nanotechnology and information technology.

best known, C_{60}, is a practically spherical structure of 60 carbon atoms; because it is hollow, it is therefore capable of holding other atoms inside. Figure 1–2 shows what one might look like—if atoms were like solid balls and you could shrink yourself down to nanoscale and still be able to see.

Fullerenes earned their discoverers the 1996 Nobel Prize in Chemistry and inspired many researchers to hunt for other remarkable structures at the nanoscale. As the Nanotech Facts webpage of the NNI notes, the development of practical applications is not automatic but can follow more or less quickly:

> The transition of nanotechnology research into manufactured products is limited today, but some products moved relatively quickly to the marketplace and already are having significant impact. For example, a new form of carbon—the nanotube—was discovered by Sumio Iijima in 1991. In 1995, it was recognized that carbon nanotubes were excellent sources of field-emitted electrons. By 2000, the "jumbotron lamp," a nanotube-based light

source that uses these field-emitted electrons to bombard a phosphor, was available as a commercial product. (Jumbotron lamps light many athletic stadiums today.) By contrast, the period of time between the modeling of the semiconducting property of germanium in 1931 and the first commercial product (the transistor radio) was 23 years.[10]

After experiencing 60 years of progress since the Manhattan Project, is nanotechnology now ready to transform the world? Encouraged by science fiction writers and visionaries who wanted to turn sci-fi dreams into reality, a romantic mythology has arisen around nanotechnology. It prophesies that nanotechnology will make practically anything possible, from cost-free manufacturing of anything humans can imagine, to cure of all diseases including old age, to extinction of the human species by self-reproducing nanoscale robotic monsters. This vision imagines that "nanotech" or "nano" will be the ultimate magic, fulfilling all human wishes and fears. As such, it has helped science fiction sustain its traditional sense of wondrous possibilities, despite widespread disappointment about the original sci-fi plot device, which was space travel to other inhabited planets.

It is good to have hope, and creative individuals need unreasonable enthusiasm to overcome the resistance of the uncreative majority and to sustain their own energies when years of effort have not led to attainment of their goals. Much nano rhetoric is hyperbole, but a certain amount of nanohype may be necessary to achieve real progress. Probably the false impressions promulgated by science fiction writers and nontechnical visionaries have helped the real scientists and engineers receive greater funding from government and industry. Perhaps they also attract young people to the related professional fields, in an era when intellectually demanding careers in science and technology are not particularly popular among the wealthy citizens of postindustrial nations like the United States. However, investors, policy makers, and interested citizens deserve an accurate accounting of the real applications that nanotechnology is likely to have.

For the United States and other advanced postindustrial societies, a crucial part of the context for nanotechnology is the heavy reliance the economy places not only on existing technology, but also on technological innovation. If the United States stops innovating, other nations with lower labor costs will take away the business that supports American prosperity. A key ingredient for innovation is entrepreneurship, but enthusiasm and salesmanship can accomplish little if science fails to provide the technical basis for innovation.

In the early 1990s, when *Scientific American* journalist John Horgan interviewed many senior scientists about whether research in their field had passed the point of diminishing returns, several of them believed that all the big discoveries had been made.[11] It should be noted that many of the scientists Horgan interviewed were very elderly, and they had an alarming tendency to die soon after he had interviewed them. Many were at the ends of their careers, if not their lives, and such people often like to think that their generation made the great discoveries and to begrudge future generations their own achievements. Even so, these scientists may have been correctly reporting that their fields, as traditionally defined, had already accomplished most of what could be expected of them.

Thus nanoconvergence may be absolutely essential for continued technological progress. The danger of hyping nanotechnology on the basis of false impressions is that its actual revolutionary potential might unfairly be discounted. A correct understanding of nanoconvergence requires serious, collaborative analysis by experts in many fields.

Technological Convergence

In a sense, nanotechnology is based on a scientific and technological convergence of great importance that began early in the twentieth century as physicists elucidated the nature of atoms. This knowledge, in conjunction with chemists' growing understanding of how atoms combined into molecules, gave birth to modern materials science. One way to understand how these fields connect is to examine how they are organized at the National Science Foundation. NSF is divided into a number of directorates, each representing a major territory of discovery. The Directorate for Mathematical and Physical Science (MPS) consists of five divisions: Mathematical Sciences, Physics, Chemistry, Materials Research, and Astronomical Sciences. We will refer to the domain of astronomical sciences in Chapter 8 (covering "the final frontier"), while the first four divisions provide the basis for most of nanoscience. The Directorate for Engineering has played a special role in organizing the National Nanotechnology Initiative in cooperation with people in MPS, other directorates, and other government agencies. Nanotechnology is not simply the current phase in the evolution of MPS fields, but rather reflects a new departure, based on their convergence, with the broadest possible implications.

The first serious effort to envision the societal implications of nanotechnology was a conference organized at the request of the Subcommittee on Nanoscale Science, Engineering, and Technology (NSET) of the U.S. government's National Science and Technology Council (NSTC), and held at NSF on September 28–29, 2000. The result was a major scientific and engineering report, *Societal Implications of Nanoscience and Nanotechnology*, edited by Mike Roco and myself. The very first sentences of the introduction to this report recognized that nanotechnology's chief impact would be through partnerships with other fields:

> A revolution is occurring in science and technology, based on the recently developed ability to measure, manipulate, and organize matter on the nanoscale—1 to 100 billionths of a meter. At the nanoscale, physics, chemistry, biology, materials science, and engineering converge toward the same principles and tools. As a result, progress in nanoscience will have very far-reaching impact.[12]

This pioneering report had great impact, both immediate and indirect. Notably, NSF began supporting projects, both large and small, to explore the social, ethical, and economic implications of nanotechnology.[13] Centers were established across the country, including the Center for Nanotechnology in Society at the University of California, Santa Barbara (grant 0531184 for $2,095,000); the Center for Nanotechnology in Society at Arizona State University (grant 0531194 for $2,605,000); and "From Laboratory to Society: Developing an Informed Approach to Nanoscale Science and Technology" associated with the nanotechnology center at the University of South Carolina (grant 0304448 for $1,350,000). A graduate research and training program was set up at MIT, "Assessing the Implications of Emerging Technologies" (grant 0333010 for $1,737,806), to involve faculty members and graduate students in prospective analysis of the likely implications of nanotechnology, based on retrospective analogies with earlier emerging technologies. The University of California, Los Angeles, began developing a database called NanoBank, providing information for social-science studies of nanoscience and commercialization (grant 0304727 for $1,490,000), specifically incorporating a component charting the convergence of nanotechnology with other fields. Finally, Michigan State University established a major convergent program called "Social and Ethical Research and Education in Agrifood Nanotechnology" (grant 0403847 for $1,720,000), with three objectives:[14]

- Deriving lessons from the social conflict over agrifood biotechnology that may be useful to the entire range of researchers engaged in the new nanotechnology initiative
- Building a new multidisciplinary competence among a team of senior researchers with extensive experience in social and ethical issues associated with agrifood technology, who have collaborated to develop communication strategies in engineering applications, and relatively junior researchers starting research programs in social and economic dimensions of agrifood science

- Identifying the most likely applications of nanotechnology within the agrifood sector (including food distribution and consumption), and developing a proactive strategy for understanding and addressing social and ethical issues associated with them

In the influential nanotechnology review called *Small Wonders, Endless Frontiers*, the National Research Council reported, "Scientists and engineers anticipate that nanoscale work will enable the development of materials and systems with dramatic new properties relevant to virtually every sector of the economy, such as medicine, telecommunications, and computers, and to areas of national interest such as homeland security."[15] Note that this sentence implies convergence, speaking of "nanoscale work" that will "enable," rather than treating nanotechnology as a completely separate branch of engineering. The NRC based its three findings about societal implications largely on our pioneering report:[16]

- The development of radically new nanotechnologies will challenge how we educate our scientists and engineers, prepare our workforce, and plan and manage R&D.

- The social and economic consequences of nanoscale science and technology promise to be diverse, difficult to anticipate, and sometimes disruptive.

- Nanoscale science and technology provides a unique opportunity for developing a fuller understanding of how technical and social systems affect each other.

As soon as we had finished editing *Societal Implications of Nanoscience and Nanotechnology*, we organized a second major gathering for December 3–4, 2001. Sponsored by NSF and the Department of Commerce, this conference examined the progress that could be achieved by combining four NBIC fields: nanotechnology, biotechnology, information technology, and cognitive science (Figure 1–3). Nearly 100 contributors concluded that this technological convergence could vastly increase the scope and effectiveness of human activity, thereby improving human performance and well-being. As co-editor Mike Roco and I explained in the first paragraph of the introduction to the report emerging from this conference:

> We stand at the threshold of a new renaissance in science and technology, based on a comprehensive understanding of the structure and behavior of matter from the nanoscale up to the most complex system yet discovered, the human brain. Unification of science based on unity in nature and its holistic investigation will lead to technological convergence and a more efficient societal

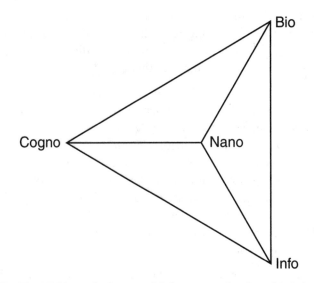

Figure 1–3 The NBIC tetrahedron combining nanotechnology, biotechnology, information technology, and new technologies based on cognitive science. Scientific and technological innovation can be stimulated through the convergence of two, three, or all four fields.

structure for reaching human goals. In the early decades of the twenty-first century, concentrated effort can bring together nano-technology, biotechnology, information technology, and new technologies based in cognitive science. With proper attention to ethical issues and societal needs, the result can be a tremendous improvement in human abilities, new industries and products, societal outcomes, and quality of life.[17]

Word-play is not a serious form of analysis, but fortuitous coincidences can express valid symbolisms. For example, NSF says it is the place where discoveries begin, based on its support for all forms of fundamental science. Thus it is not surprising that NSF supported the first NBIC conference. NBIC will transform the world, and the letters "NSF" are at the heart of the word "traNSForm"! Converging technologies seek to combine the powers of all sciences, and the letters "NBIC" are found in "ComBINe." They are also found in "BioNIC," the combination of biotechnology and information technology to enhance human performance.

Many perceptive observers have noticed the progressing convergence. In his massive study of the Information Society, Manuel Castells writes, "Technological convergence increasingly extends to growing interdependence between the biological and micro-electronics revolutions, both materially and methodologically. . . . Nanotechnology may allow sending tiny microprocessors into the systems of living organisms, including humans."[18] Leading scientists have actively promoted convergence throughout their careers—most notably sociobiologist Edward O. Wilson, who called convergence "consilience" in his 1998 book of that title.[19]

The challenge of integrating fields, disciplines, and subdisciplines will stimulate both theoretical creativity and empirical discovery. Measurement techniques developed in one area will accelerate progress elsewhere, as will innovative tools of all kinds, from nanoscale sensors to cyberinfrastructure. Investment by government and industry cannot be entirely justified by the anticipated intellectual benefits, however. The great promise of technological convergence must attract the interest of policy makers and ordinary citizens through the practical applications it can achieve. Converging technologies will make people healthier, stronger, smarter, more creative, and more secure. In their group deliberations and individual essays, the prominent scientists and engineers at the NBIC conference identified a variety of practical possibilities associated with this trend:[20]

- Comfortable, wearable sensors and computers will enhance every person's awareness of his or her health condition, environment, chemical pollutants, potential hazards, and information of interest about local businesses, natural resources, and the like.

- Machines and structures of all kinds, from homes to aircraft, will be constructed of materials that have exactly the desired properties, including the ability to adapt to changing situations, high energy efficiency, and environmental friendliness.

- A combination of technologies and treatments will compensate for many physical and mental disabilities and will eradicate altogether some handicaps that have plagued the lives of millions of people.

- Robots and software agents will be far more useful for people, because they will operate on principles compatible with human goals, awareness, and personality.

- People from all backgrounds and of all ranges of ability will learn valuable new knowledge and skills more reliably and quickly, whether in school, on the job, or at home.

- Individuals and teams will be able to communicate and cooperate profitably across traditional barriers of culture, language, distance, and professional specialization, thereby greatly increasing the effectiveness of groups, organizations, and multinational partnerships.

- The human body will be more durable, healthier, more energetic, easier to repair, and more resistant to many kinds of stress, biological threats, and aging processes.

- National security will be greatly strengthened by lightweight, information-rich war-fighting systems, capable uninhabited combat vehicles, adaptable smart materials, invulnerable data networks, superior intelligence-gathering systems, and effective measures against biological, chemical, radiological, and nuclear attacks.

- Anywhere in the world, an individual will have instantaneous access to needed information, whether practical or scientific in nature, in a form tailored for most effective use by that particular individual.

- Engineers, artists, architects, and designers will experience tremendously expanded creative abilities, both with a variety of new tools and through improved understanding of the wellsprings of human creativity.

- The ability to control the genetics of humans, animals, and agricultural plants will greatly benefit human welfare; widespread consensus about ethical, legal, and moral issues will be built in the process.

- The vast promise of outer space will finally be realized by means of efficient launch vehicles, robotic construction of extraterrestrial bases, and profitable exploitation of the resources of the Moon, Mars, or near-Earth-approaching asteroids.

- New organizational structures and management principles based on fast, reliable communication of needed information will vastly increase the effectiveness of administrators in business, education, and government.

- Both average persons and policy makers will have a vastly improved awareness of the cognitive, social, and biological forces operating their lives, enabling far better adjustment, creativity, and daily decision making.

- The factories of tomorrow will be organized around converging technologies and increased human–machine capabilities as "intelligent environments" that achieve the maximum benefits of both mass production and custom design.

- Agriculture and the food industry will greatly increase yields and reduce spoilage through networks of cheap, smart sensors that constantly monitor the condition and needs of plants, animals, and farm products.

- Transportation will be safe, cheap, and fast owing to ubiquitous real-time information systems, extremely high-efficiency vehicle designs, and the use of synthetic materials and machines fabricated from the nanoscale for optimal performance.

- The work of scientists will be revolutionized by importing approaches pioneered in other sciences—for example, genetic research employing principles from natural language processing and cultural research employing principles from genetics.

- Formal education will be transformed by a unified but diverse curriculum based on a comprehensive, hierarchical intellectual paradigm for understanding the architecture of the physical world from the nanoscale through the cosmic scale.

- Fast, broadband interfaces directly between the human brain and machines could transform work in factories, control automobiles, ensure military superiority, and enable new sports, art forms, and modes of interaction between people.

Since the original Converging Technologies conference, there have been three others—in Los Angeles, New York, and Kona, Hawaii—plus a second NSF-organized conference on the societal implications of nanotechnology that confirmed the centrality of nanoscience for convergence, and of convergence for the impacts of nanotechnology. I had the privilege of co-editing five of the book-length reports that grew out of these conferences and contributing two chapters to the sixth report; I also had the pleasure of attending all of these historic gatherings. In addition, the European Commission (EC) published a report in reaction to the U.S. work in this field; the EC report, called *Converging Technologies: Shaping the Future of European Societies*, urged concerted efforts in this area.[21]

Application Areas

At the first Converging Technologies conference, five workshop groups of experts in appropriate fields considered the research challenges associated with highly valuable applications that could enhance human performance along five different dimensions. Their conclusions follow.[22]

Expanding Human Cognition and Communication. The human mind can be significantly enhanced through technologically augmented cognition, perception, and communication. Central to this vital work will be a multidisciplinary effort to understand the structure and function of the mind, which means research not only on the brain, but also on the ambient sociocultural milieu, which both shapes and is shaped by individual thought and behavior. Specific application areas include personal sensory device interfaces and enhanced tools for creativity. A fundamental principle is putting people fully in command of their technology, which will require sociotechnical design to humanize computers, robots, and information systems.

Improving Human Health and Physical Capabilities. In the absence of new approaches, medical progress is widely expected to slow markedly during the coming century. To increase longevity and well-being throughout the life span, we will need to innovate in fresh areas. Nanoscale biosensors and bio-processors can contribute greatly to research and to development of treatments, including those resulting from bioinformatics, genomics, and proteomics. Implants based on nanotechnology and regenerative biosystems may replace human organs, and nanoscale machines might unobtrusively accomplish needed medical interventions. Advances in cognitive science will provide insights to help people avoid unhealthy lifestyles, and information technology can create virtual environment tools both for training medical professionals and for enlisting patients as effective partners in their own cure.

Enhancing Group and Societal Outcomes. Peace and economic progress require vastly improved cooperation in schools, corporations, government agencies, communities, and nations, as well as across the globe. Unfortunately, communication is too often blocked by substantial barriers caused by physical disabilities, language differences, geographic distances, and variations in knowledge. These barriers can be overcome through the convergence of cognitive and information science to build a ubiquitous, universal web of knowledge, which is automatically translated into the language and presentation media desired by diverse users. Nano-enabled microscale data devices will identify every product and place, and individuals will merge their personal databases as they choose which groups and interaction networks to join. Group productivity tools will radically enhance the ability of people to imagine and create revolutionary new products and services based on the integration of the four technologies from the nanoscale.

National Security. The rapidly changing nature of international conflict demands radical innovations in defense technology, strategic thinking, and the capabilities of professional war fighters. Both mental and physical enhancement of human abilities can achieve significant gains in the performance of individual military personnel, and new battlefield communication systems employing data linkage and threat anticipation algorithms will strengthen armies and fleets. The combination of nanotechnology and information technology will produce sensor nets that are capable of instantly detecting chemical, biological, radiological, and explosive threats and can direct immediate and effective countermeasures. Uninhabited combat vehicles and human–machine interfaces will enhance both attack capabilities and survivability. As was true historically in the development of computer technology, developments initially achieved at high cost for defense purposes will be transferred over time to low-cost civilian applications, for the general benefit of society.

Unifying Science and Education. To meet the coming challenges, scientific education needs radical transformation at all stages, from elementary school through postgraduate training. Convergence of previously separate scientific disciplines and fields of engineering cannot take place without the emergence of new kinds of people who understand multiple fields in depth and can intelligently work to integrate them. New curricula, new concepts to provide intellectual coherence, and new forms of educational institutions will be necessary.

Radical Transformations

Revolutionary advances at the interfaces between previously separate fields of science and technology are ready to create key *transforming tools* for NBIC technologies. These tools include scientific instruments, analytical methodologies, radically new materials, and data-sharing systems. The innovative momentum achieved in these interdisciplinary areas must not be lost, but rather should be harnessed to accelerate unification of the various disciplines. Progress can become self-catalyzing if we press forward aggressively; if we hesitate, however, the barriers to progress may crystallize and become harder to surmount.

Developments in systems approaches, mathematics, and computation in conjunction with NBIC allow us for the first time to understand the natural world, human society, and scientific research as closely coupled, complex, hierarchical systems. At this moment in the evolution of technical achievement, improvement of human performance through integration of technologies becomes possible. When applied both to particular research problems and to the overall organization of the research enterprise, this complex systems approach provides holistic awareness of opportunities for integration, thereby allowing us to obtain the maximum synergy along the main directions of progress.

One reason sciences have not merged in the past is that their subject matter is so intellectually complex. It will often be possible to rearrange and connect scientific findings, based on principles from cognitive science and information theory, so that scientists from a wider range of fields can comprehend and apply those findings within their own work. Researchers and theorists must look for promising areas in which concepts developed in one science can be translated effectively for use in another science. For example, computational principles developed in natural language processing can be applied to work in genomics and proteomics, and principles from evolutionary biology can be applied to the study of human culture.

The aim of NBIC convergence is to offer individuals and groups an increased range of attractive choices while preserving such fundamental values as privacy, safety, and moral responsibility. It can give us the means to deal successfully with the often unexpected challenges of the modern world by substantially enhancing our mental, physical, and social abilities. Most people want to be healthier and to live longer. Most people want prosperity, security, and creativity. By improving the performance of all humans, technological convergence can help all of us achieve these goals together.

As the challenge posed by national security illustrates, human performance is often competitive in nature. In this arena, what may matter is the *relative* military power of two contending armies or the *relative* economic power of two competing corporations, not their *absolute* power. At the present time, technologically advanced nations such as the United States, Japan, and the countries of Western Europe maintain their positions in the world order in significant part through their rate of technical progress. Conversely, "developing countries" provide raw materials and relatively low-tech manufactured commodities in exchange for the cutting-edge products and services that the advanced nations can offer. If a rich nation were to cease moving forward technologically, a much poorer nation could quickly match the quality of its exports at lower cost. Although this reversal of fortune would be fine for businesses in the poorer nation, the rich nation could see its standard of living drop rapidly toward the world average. The result in such a case might be not merely disappointment and frustration, but deep social unrest.

For example, a significant fraction of the prosperity of the United States depends on the continuing superiority of its information technology, including the components manufactured by its semiconductor industry. In 1965, Gordon Moore, the co-founder of the Intel Corporation, observed that the density of transistors on the most advanced microchip doubles about every 18 months. Dubbed *Moore's law*, this observation has proven to be true ever since. Now, however, the transistors on conventional chips are nearing physical size limits that could repeal this "law" within a decade. If that happens, the U.S. semiconductor industry may evaporate, as other nations catch up to the current U.S. technical lead and produce comparable chips at lower cost. Not surprisingly, both U.S. government and industry have recently developed intense interest in nanotechnology approaches that could potentially extend the life of Moore's law by another decade or two—most notably, molecular logic gates and carbon nanotube transistors.

The realization of these radically new approaches will require the development of an entire complex of fresh technologies and supporting industries, so the cost of shifting over to them may be huge. Only the emergence of a host of new applications could justify the massive investments, by both government and industry, that will be required to make this transition. Already, there is talk in the computer industry of "performance overhang"—that is, the possibility that technical capabilities have already outstripped the needs of desirable applications. For example, the latest models of home computers are finally able to handle the speed and memory demands of high-quality video, but no more-demanding application is currently on the horizon that would require a new generation of hardware.

During the twentieth century, several major technologies essentially reached maturity or ran into social, political, or economic barriers to progress. Aircraft and automobiles, for example, have changed little in recent years. The introduction of high-definition television has been painfully slow, and one would predict that consumers will be content to stick with the next generation of television sets for many years. The evolution of spaceflight technology has apparently stalled at about the technical level of the 1970s, and the advance of nuclear technology has either halted or been blocked by political opposition. In medicine, the rate of introduction of new drugs has slowed, and the great potential of genetic engineering is threatened by increasing popular hostility. In short, technological civilization faces the very real danger of stasis or decline unless something can rejuvenate progress.

The *Converging Technologies* report suggests that the unification of nanotechnology, biotechnology, information technology, and cognitive science could launch a New Renaissance. Five centuries ago, the Renaissance energized all fields of creative endeavor by infusing them with the same holistic spirit and shared intellectual principles. It is time to rekindle the spirit of the Renaissance, returning to the holistic perspective on a higher level, with a new set of principles. In the first Renaissance, a very few individuals could span multiple fields of productivity and become "Renaissance men." Today, technological convergence holds out the very real hope that all people on the planet could become "Renaissance people" by taking advantage of enhanced abilities, tools, materials, knowledge, and humane institutions.

THE PLAN OF THIS BOOK

This chapter has reported the conclusions of the scores of leading scientists who participated in the Societal Implications and Converging Technologies workshops: Nanoscience and nanotechnology will have immense implications for human society. Although nano will generate distinctive materials and products, its chief impact will be felt through collaboration with other fields. Convergence at the nanoscale will unite nanotechnology with biotechnology, information technology, and new technologies based on cognitive science. Without this unification, scientific, technological, and economic progress would be greatly in doubt. This chapter has also described some of the research carried out at the nanoscale and hinted at likely applications of nanoconvergence that may emerge over the coming decade or two.

Chapter 2 deals with the fantasies and illusions that have both popularized the nano concept and given many investors, policy makers, and ordinary citizens a seriously distorted picture of the field. We cannot properly understand how nanotechnology will converge with the other fields if we have a false impression of the field itself. Also, nanofantasies would prevent us from seeing the real importance of convergence at the nanoscale, because we would falsely imagine that nano alone would remake the world without need of all the rest of the sciences and technologies. Chapter 2 uses a pair of parables plus sci-fi storytelling to show how science fiction literature has long promulgated inspiring but factually false impressions of the nanoscale. Some of these illusions involve convergence, especially Eric Drexler's original conception that nanotechnology is mechanical engineering applied to chemistry on the molecular scale.

Chapter 3 focuses on information technology and its convergence with nanotechnology. Already, the smallest transistors on computer chips are less than 100 nanometers across, and hard-disk memory storage exploits nanoscale magnetic phenomena. Moore's law has driven progress across all domains of information technology, but we may have reached the point at which this decades-long period of computer chip performance progress comes to a close, unless nanotechnology can take us further. Other promising areas of research, notably in nano-enabled microscale sensors and in quantum computing, could benefit from progress in nanotechnology. At the same time, information technology contributes directly to progress in all fields of science and engineering, and we may have entered a period in which the most important tool of research and development is cyberinfrastructure.

Chapter 4 focuses on the interface between nanotechnology and biotechnology, a tremendously active area of research at the present time. Both the National Institutes of Health (NIH) and NSF have aggressively supported research in nanobiotechnology (also known as bionanotechnology). The fundamental structures inside living cells that do all the work of metabolism and reproduction are nanoscale "machines" composed of complex molecular structures, and the methods of nanoscience are needed to understand them. For a century, biologists and medical researchers have sought to solve the problem of cancer, and nanobioconvergence offers new hope that this effort will finally succeed. Concepts from biology have been applied to information technology, and new biotechnologies enabled by both nano and info promise to improve human physical and mental performance.

Cognitive science, the subject of Chapter 5, is itself a convergence of disciplines, combining artificial intelligence, linguistics, psychology, philosophy, neuroscience, anthropology, and education. "Cog-sci" was initially dominated by the paradigm espoused by classical artificial intelligence, which modeled human thought processes in terms of logical manipulations of clearly defined, high-level concepts. More recently, a wide range of other paradigms have been introduced by this field's convergence with other branches of information technology and with biotechnology and nanotechnology. Society faces a number of challenges if it is to digest the cultural implications of cognitive science, notably the emerging controversies about the future viability of religion and neurotechnologies that could transform human cognition. An NBIC task force suggested that the greatest near-term development coming out of a union of cognitive science with other fields would be an information technology system, called *The Communicator*, that might transform human interaction.

Chapter 6 considers how we could accomplish full convergence of the NBIC fields as well as their convergence with reformulated social sciences. I suggest a system of theoretical principles—conservation, indecision, configuration, interaction, variation, evolution, information, and cognition—that could help connect similar natural laws, research methods, and technological applications across all these fields. Policy decisions about investment in various technologies require serious consideration of the ethical principles at stake and the likely social effects of those decisions. However, we cannot examine those issues rigorously without benefit of social science, and many knowledgeable people doubt that the social sciences are equal to the task, at least as currently constituted. To illustrate this crucial point, Chapter 6 describes a linked pair of failed attempts to accomplish convergence across the social sciences half a century ago, coming to the ironic conclusion that both were headed in the right direction but premature. A fresh attempt to unify and strengthen the social sciences could succeed, if it were based on solid cognitive science in convergence with the other NBIC fields.

Chapter 7 acknowledges that the social sciences cannot give us definitive answers to vital questions at the present time, but collects together a wealth of ideas about how convergence might affect human society. Already having surveyed the views of scientists in earlier chapters, we consider the harshest critics of convergence and the notions of ordinary citizens about the future of the world. A dozen years ago, social scientists proposed a major initiative to strengthen their disciplines so as to better understand the nature of our rapidly changing world, and it is not too late to follow their advice. More recently, key participants in the convergence movement have urged the creation of a new branch of social science focusing on service industries, an idea that is fully compatible with the decade-delayed hope to develop a convergent science of democratic institutions. Standing still is not an option, because uncontrolled sociopolitical forces will harness new technologies to divergent forces ripping humanity apart. The only hope is unification of the world on the basis of the unification of science.

The final chapter offers a visionary but scientifically based vision of how nanoconvergence might transform human potentialities by enabling vigorous exploration and colonization of outer space. Although convergence has vast terrestrial implications, it is easier to see clearly how NBIC fields could combine to create a revolution in astronautics. Specifically, they could revolutionize human access to the solar system, thereby leading to exploitation of the environments and resources that exist beyond the Earth. Current technologies are not potent enough to build an interplanetary society. By enabling moderate improvements across all space-related technologies, however, nanoconvergence could potentially help humans enter the final frontier with the powers needed to accomplish previously unimaginable goals. On new worlds, we could reinvent ourselves, our society, and our destiny.

REFERENCES

1. Norio Taniguchi, "On the Basic Concept of 'Nano-Technology'," *Proceedings of the International Conference Production Engineering, Tokyo*, Part II, Japan Society of Precision Engineering, 1974.

2. William Sims Bainbridge, "Sociocultural Meanings of Nanotechnology: Research Methodologies," *Journal of Nanoparticle Research*, 6:285–299, 2004.

3. Mihail C. Roco, "The Action Plan of the U.S. National Nanotechnology Initiative," in Mihail C. Roco and Renzo Tomellini (eds.), *Nanotechnology: Revolutionary Opportunities and Societal Implications* (Brussels, Belgium: European Commission, 2002, p. 31).

4. Jennet Conant, *Tuxedo Park: A Wall Street Tycoon and the Secret Palace of Science That Changed the Course of World War II* (New York: Simon and Schuster, 2002).

5. Henry De Wolf Smyth, *Atomic Energy for Military Purposes* (Princeton, NJ: Princeton University Press, 1945).

6. http://www.nano.gov/html/facts/home_facts.html

7. Ernst Ruska, "The Development of the Electron Microscope and of Electron Microscopy," in Tore Frängsmyr and Gösta Ekspång (eds.), *Nobel Lectures, Physics 1981–1990* (Singapore: World Scientific Publishing, 1993, pp. 355–380).

8. Gerd Binnig and Heinrich Rohrer, "Scanning Tunneling Microscopy: From Birth to Adolescence," in Tore Frängsmyr and Gösta Ekspång (eds.), *Nobel Lectures, Physics 1981–1990* (Singapore: World Scientific Publishing, 1993, p. 392).

9. Robert F. Curl, Jr., "Dawn of the Fullerenes: Experiment and Conjecture," in Ingmar Grenthe (ed.), *Nobel Lectures, Chemistry 1996–2000* (Singapore: World Scientific Publishing, 2003, pp. 11–32); Harold Kroto, "Symmetry, Space, Stars and C60," in Ingmar Grenthe (ed.), *Nobel Lectures, Chemistry 1996–2000* (Singapore: World Scientific Publishing, 2003, pp. 44–79); Richard E. Smalley, "Discovering the Fullerenes," in Ingmar Grenthe (ed.), *Nobel Lectures, Chemistry 1996–2000* (Singapore: World Scientific Publishing, 2003, pp. 89–103).

10. http://www.nano.gov/html/facts/home_facts.html

11. John Horgan, *The End of Science: Facing the Limits of Knowledge in the Twilight of the Scientific Age* (Reading, MA: Addison-Wesley, 1996).

12. Mihail C. Roco and William Sims Bainbridge (eds.), *Societal Implications of Nanoscience and Nanotechnology* (Dordrecht, Netherlands: Kluwer, 2001, p. 1).

13. http://www.nsf.gov/awardsearch/

14. http://www.nsf.gov/awardsearch/showAward.do?AwardNumber=0403847

15. National Research Council, *Small Wonders, Endless Frontier: A Review of the National Nanotechnology Initiative* (Washington, DC: National Academy Press, 2002, p. 1).

16. National Research Council, *Small Wonders, Endless Frontier: A Review of the National Nanotechnology Initiative* (Washington, DC: National Academy Press, 2002, pp. 31–32).

17. Mihail C. Roco and William Sims Bainbridge, "Overview: Converging Technologies for Improving Human Performance," in Mihail C. Roco and William Sims Bainbridge (eds.), *Converging Technologies for Improving Human Performance* (Dordrecht, Netherlands: Kluwer, 2003, p. 1).

18. Manuel Castells, *The Rise of the Network Society* (Oxford, UK: Blackwell, 2000, p. 72).

19. Edward O. Wilson, *Consilience: The Unity of Knowledge* (Thorndike, ME: Thorndike Press, 1998); Ullica Segerstrale, "Wilson and the Unification of Science," in William Sims Bainbridge and Mihail C. Roco (eds.), *Progress in Convergence* (New York: New York Academy of Sciences, 2006, pp. 46–73).

20. Mihail C. Roco and William Sims Bainbridge, "Overview: Converging Technologies for Improving Human Performance," in Mihail C. Roco and William Sims Bainbridge (eds.), *Converging Technologies for Improving Human Performance* (Dordrecht, Netherlands: Kluwer, 2003, pp. 5–6).

21. Alfred Nordmann (ed.), *Converging Technologies: Shaping the Future of European Societies* (Brussels, Belgium: European Commission, 2004), http://ec.europa.eu/research/conferences/2004/ntw/pdf/final_report_en.pdf

22. William Sims Bainbridge, "Converging Technologies (NBIC)," in *Nanotech 2003: Technical Proceedings of the 2003 Nanotechnology Conference and Trade Show* (Boston: Computational Publications, 2003, pp. 389–391).

Chapter 2

Visions and Illusions

Science fiction writers and nanotech propagandists have spread false rumors about the probable impact of nanotechnology, such as the notion that self-reproducing nanobots could soon either achieve cost-free manufacturing or threaten all life on Earth. In truth, as other chapters of this book show, the real potential of nanotechnology is nearly as remarkable and lies chiefly in its implications for other technologies. This chapter examines both the visionaries and their visions. Nanotechnology offers many realistic investment opportunities, plus perhaps a few dangers, but finding them is made more difficult by the unrealistic nanohype currently being spread by fiction writers and undisciplined visionaries. Of course, visionaries do deserve to be honored when they offer new ideas, fresh perspectives, and transcendent goals.

IMAGINATION AND IMPOSSIBILITY

My father used to say, "What man can imagine, man can do." Like many who came to adulthood before the disillusionments of recent years, he believed that the rapid technological development he had seen during his lifetime would continue indefinitely and greatly benefit humanity. A top executive of a major life insurance company, he was not so naive as to believe that nothing was impossible. Rather, my father thoroughly understood the work of the actuaries in his company, who constantly updated estimates of the risks faced by policy holders and the company itself. He also understood the management of the company's investments and the necessity of having a sufficient rate of return and sales of new policies to offset payments to beneficiaries of the life insurance policies. That is, he knew that the iron-clad laws of economics and demography determined the success of his company, so it would be futile to imagine a new management approach that ignored those realities.

For scientists, engineers, and businesspeople, impossibilities are the things that give you leverage to exploit possibilities. Technology exploits the

Figure 2–1 "Dr. Nano." This is a "nanograph" portrait of nanotechnology and convergence leader Mihail C. Roco, created at Oak Ridge National Laboratory. The individual pixels of this image are about 50 nanometers in diameter, and the face is about 3,000 nanometers high, so the entire picture is far too small to be seen by the unaided human eye. This work of art symbolizes the importance of imagination and vision, especially when disciplined by scientific fact.

laws of nature, rather than violating them. The facts set real limits on what we can do, among all the things we can imagine. Further insight can be found in two parables: "Birds Can't Fly to the Moon" and "Cold Facts."

Birds Can't Fly to the Moon

Many millions of years ago, birds were chicken-sized dinosaurs, hopping and pecking around in the forest, the first parable says. They possessed feathers, which they imagined existed because they were so beautiful but probably just kept the birds warm like the fur on the crawly mammals they disdained. Over the generations, their feathered arms changed shape, perhaps to help them keep balance as they ran or to control their leaps from limb to limb in the abundant trees.

One night, a particularly poetic bird named Protoavis saw a beautiful light in the sky and decided to jump to it, not knowing that it was the Moon. Straining to reach the light, he flapped his arms and was able to fly almost level with the ground for 100 yards before falling. The next morning, in great excitement, Protoavis recounted his exploit to his fellows, and they found that with some effort they could fly short distances, too.

Over many further generations, birds flew higher and better, until they were zooming high in the sky. They recounted the legend of Protoavis, and predicted that one day a bird would fly high enough to reach the Moon. Rumors suggested that a great condor in the Andes had already flown over the highest mountain. All of the little birds were taught that soon—maybe in their own generation—a bird would fly to the Moon. But it never happened, because the lunar optimism of the birds was misplaced.

Three kinds of limits prevented birds from flying to the Moon: practical, absolute, and strategic.

Flying very high was not an evolutionary advantage, because it does not help any bird survive or reproduce. Flying just a few hundred feet is enough to escape any jumping cat, and there is no food in the stratosphere. Thus there is no practical reason for a bird to fly above the clouds.

The absolute limit is the fact that there is no air between the Earth and the Moon, so wings cannot work in this environment. So long as birds fly with wings, the voyage is strictly impossible.

Wings could metaphorically be described as a strategic choice. Of course, the choice was taken by evolution in a process of natural selection, rather than by vote of a parliament of fowls. An alternative choice would have been to develop hands into tools for manipulating objects, but instead the birds used their finger bones for the leading edges of their wings. There is a way to fly to the Moon, but it requires intelligence and centuries of advancing technology to build space rockets. Creatures with hands did that, not creatures with wings.

There are two ways to apply this parable to humans. First, we can seek parallels for each of the analytic ideas. For humans, practical limits might be economic rather than evolutionary. Absolute limits are suggested by our scientific knowledgebase: no perpetual motion, no travel faster than light, and no free lunch. Strategic limits exist when we preclude certain choices by making incompatible choices, even though we are often not aware of the alternative possibilities. I have often wondered why the classical civilizations of Greece and Rome stopped innovating and fell. One possible explanation is that their reliance upon slave labor gave the people who were working with technology neither motivation nor permission to experiment with new ways. That was a bad strategic decision, as well as bad ethics.

A second way to apply the parable is as a whole, by seeing a general analogy in humans. Perhaps our intelligence is like birds' wings. It evolved in our originally unimportant species in such a way as to give us an advantage, on the basis of random changes in our habitat that had shaped us to be ready to exploit our bigger brains—for example, through making things with our hands. Innovation fed innovation, until here we are with tremendous capabilities but wondering how much farther we can go. It is quite possible we have nearly completed exploiting the new ecological niche opened to us by intelligence. Birds have reached the limits of their adaptive radiation; perhaps we humans have nearly reached ours.

Cold Facts

The second parable concerns an invention I would like to bring to fruition. I have the basic idea, and I am sure all the rest will just be minor details. I call my invention the Cooler. It is the size of a pack of playing cards or a pocket music player. There is a dial on the side, which I can set to any desired temperature. If I put the Cooler in a Styrofoam container of soft drinks and set it for just above freezing, it will keep the drinks refreshingly cold. If the bedroom gets hot, I set the Cooler for a lower temperature and slumber in cool peace.

How does the Cooler work, you may ask? Well, it neutralizes the heat. As you know, heat is just vibrating molecules, so the Cooler takes molecules that just happen to be vibrating in opposite directions and makes them cancel out.

Did you say that the Cooler violates the conservation of energy? Refrigerators, you point out, radiate the heat away from the coils in back or dispose of the heat by some other means. So where does the Cooler's heat go?

Oh, I forgot to mention that the Cooler uses nanotechnology. As everyone knows, Einstein showed that matter and energy can be transformed into each other. Inside the Cooler, then, little nanotechnology fingers take puffs of energy and turn them into nanoscale dust motes of matter. You can imagine the process as being like a person in a snowstorm who grabs falling flakes of snow and presses them into snowballs. Because vast amounts of energy make very little matter, the dust collects in a compartment inside the Cooler that will not fill up for a century, by which time you can trade it in for a new model.

I'll end the parable right here, so I won't have to tell you how the nanotechnology accomplishes such things. I would probably mutter something about how physicists have shown that gamma rays can collide to produce an electron and a positron, thereby transmuting energy into mass. You won't be allowed to ask troubling questions such as how the low-energy photons of heat can duplicate what only the high-energy photons of gamma rays can do, where all the positrons go so they won't smash into electrons and revert to gamma rays, or where I get the quarks to make the protons needed for my dust motes. The magic word "nanotechnology" silences all those questions.

The "Cold Facts" parable satirizes qualitative, analogical thinking of the kind that makes for good science fiction stories. Granting a couple of impossible assumptions, you weave a yarn that superficially seems logical, using an entertaining story to divert the reader so he or she does not check your calculations.

How do you store the universal solvent, an acid that will dissolve anything? You make a cup by freezing the solvent itself, and keep it in that. How do you travel through time? You recognize that time is the fourth dimension, so all you need to do is find the right direction in which to go. How do you travel faster than light? Velocity is distance divided by time, so you reduce the distance by warping space. These ideas from science fiction are certainly interesting, they exercise the minds of young people, and they inspire real scientists and engineers to achieve. But they are not realistic.

One of the most widely acclaimed science fiction stories, "The Cold Equations" by Tom Godwin, critiques the technological optimism of this literature.[1] An innocent young girl must be jettisoned into airless space without the slightest hope of survival, simply because the cold equations of astronautics dictate that the rocket on which she has stowed away will crash if forced to carry her extra weight. Parables of cold facts and cold equations put science fiction in perspective.

SCIENCE FICTION

There exists an old—if minor—tradition in science fiction that sets stories in very tiny environments. Way back in 1919, the pages of the pulp fiction magazine *All-Story Weekly* carried a marvelous tale by Ray Cummings, "The Girl in the Golden Atom." Later expanded into a novel, this story concerned a scientist who was able to shrink himself down far below the nanoscale to visit an electron, which was depicted in a manner akin to a planet circling an atomic nucleus as its sun. This adventure begins when the scientist focuses a powerful new microscope on the wedding ring of his mother and sees an attractive female inhabitant of the atomic planet in great danger. He has the technology to zoom down and rescue her, but if his instruments ever lose track of that particular atom, he can never find it again.

Cummings actually was well informed about science and technology, because he was an assistant of the great electrical genius, Thomas Alva Edison. In 1911, physicist Ernest Rutherford had discovered that atoms consisted of a nucleus surrounded by electrons, and his "solar system" model of the atom received considerable publicity. Rutherford had used gold foil as the target of his experiments, which naturally suggested a gold wedding ring for Cummings' romance. Even in 1911, it was clear that the structure of atoms was very different from the structure of our solar system, however. For example, electron orbits are not limited roughly to a plane, as is true for the planets, and several identical electrons can exist in the same orbit. In 1913, Niels Bohr suggested that electron orbits are possible only at certain energies, a concept that led to the quantum theory developed over the next few years. Today, the modern scientific picture of the atom is very different from a solar system. Importantly, electrons and other fundamental constituents of matter cannot function as worlds in which intelligent creatures could live and have adventures or romances, because below a certain size (called the Planck length) or energy level (called the quantum) no stable structures are possible.[2]

This is not to say that Cummings should not have written his entertaining story. It probably encouraged readers who happened to be teenage boys to believe that atoms were interesting, and when the novel came out in 1923 its science was only a decade out of date. Later writers borrowed the idea that electrons could be valid settings for stories—see, for example, Festus Pragnell's *The Green Man of Graypec*. In the critically acclaimed story "Surface Tension," James Blish imagined aquatic people only a millimeter in height, whose cells would of necessity be nanoscale, struggling to break through the conservatism of their water-bound culture, for which surface tension is a nice metaphor.[3]

Perhaps the best-known story about adventures at the nanoscale is *Fantastic Voyage* by Isaac Asimov, novelized from a 1966 movie with the same title starring Raquel Welch and Stephen Boyd, in which a submarine and its crew are miniaturized and injected into the bloodstream of a human being to destroy a dangerous blood clot in his brain.[4] If you were to try to miniaturize people to this degree, you would have to choose between two ways to do it: Would you remove many of the humans' atoms, until just enough were left to fill the space available, or would you shrink every single atom in the humans' bodies? Imagine you had some magical tweezers that could reach in and pluck out every other atom or every other cell. Before too long, the person would begin to lose memories and vital functions, and there would not be enough left to operate a submarine long before you had made the crew small enough to inject through a hypodermic needle. If you chose the other approach, shrinking each atom, you would immediately run into the problem that the fundamental constituents of atoms have unalterable sizes. Every electron, for example, has exactly the same rest mass and other characteristics as every other electron in the universe.

Asimov, who was trained as a biochemist, was one of the leading writers of the "hard science" variety of science fiction that sought to craft interesting stories based on speculations in science and technology, without violating more than one or two natural laws per story. Like Sir Isaac Newton, Asimov invented three laws, but they concerned technology, rather than science. That is, Asimov's Three Laws of Robotics specify how robots should be programmed so that they behave properly. Although these laws are most assuredly famous, they have had no influence whatsoever within the real discipline of robotics. Indeed, as Asimov's own stories illustrate, it may be impossible to reduce morality to simple principles like his First Law: "A robot may not injure a human being or, through inaction, allow a human being to come to harm."[5] As philosopher Daniel C. Dennett has observed, "No remotely compelling system of ethics has ever been made computationally tractable, even indirectly, for real-world problems."[6]

Asimov and his colleague Arthur C. Clarke wrote both science fiction and factual popularized science. While riding together in a taxi cab, the two of them concluded a treaty, specifying that one was the greatest science fiction writer in the world and the other was the greatest writer of popular science. Unfortunately, they could not agree on which person deserved each sobriquet.[7] This is a nice metaphor for the problem of deciding how much fact versus how much fiction exists in anything written about science for a popular audience. Like Newton and Asimov before him, Clarke also formulated three laws:[8]

1. When a distinguished but elderly scientist states that something is possible, he is almost certainly right. When he states that something is impossible, he is very probably wrong.

2. The only way of discovering the limits of the possible is to venture a little way past them into the impossible.

3. Any sufficiently advanced technology is indistinguishable from magic.

These poetic, ironic, appealing principles belong to science fiction, not to science fact. If the first principle were true, then anything about which a distinguished but elderly scientist might have an opinion is very probably true. The second is logically contradictory, unless one adds the qualifier "in the imagination." The third suggests that the technology in science fiction serves the same function that magic does in fantasy fiction, permitting emotionally satisfying stories that otherwise would contradict our knowledge of the real world. A historian of the genre, Sam Moskowitz, explains, "Science fiction is a branch of fantasy identifiable by the fact that it eases the 'willing suspension of disbelief' on the part of its readers by utilizing an atmosphere of scientific credibility for its imaginative speculation in physical science, space, time, social science, and philosophy."[9]

To the general public, perhaps the most familiar sci-fi novel about nanotechnology is *Prey* by Michael Crichton, the famous author of *Jurassic Park* and a host of other novels that depict science as irresponsible (and have been made into blockbuster movies).[10] Crichton's introduction warns, "Sometime in the twenty-first century, our self-deluded recklessness will collide with our growing technological power. One area where this will occur is in the meeting point of nanotechnology, biotechnology, and computer technology. What all three have in common is the ability to release self-replicating entities into the environment."[11] Practicing chemists and engineers in the relevant fields doubt that engineered nanoscale entities could become self-reproducing in the natural environment, unless they were genetically engineered on the basis of the living things that have co-evolved with the Earth. Crichton's novel postulates that a monster might be created by a corporate research project gone awry, with the dangerous entity consisting of biologically launched nanoparticles that evolve rapidly and develop a swarm intelligence. In the first of two climaxes, the monster kills people; in the second, it takes control over their bodies.

Prey may be sci-fi or mainstream horror, but it is not really science fiction as classically defined, because it lacks the sense of wonder that marks this genre of literature.[12] In contrast, Greg Bear's *Blood Music* qualifies as real science fiction because it explores a variety of intellectually interesting, transcendent possibilities that might conceivably result from a breakthrough in nanotechnology.[13] In the original 1983 novelette, a scientist is working on a corporate project to develop molecular computing devices on the nanoscale, through convergence of computer chips with genetic engineering. After a dispute with his company, he injects the result into his own circulatory system. The devices multiply and evolve into nanoscale intelligences that first modify the scientist's body, and then consciously develop means to infect (or, from their perspective, colonize) other people. The story ends as the bodies of a doctor and his wife merge into a single organism, in symbiosis with the nanoscale civilization living within them. Subsequently, Bear expanded the story into a novel that describes the fantastic transformation of all humanity. Whereas *Prey* is simply a crude monster fantasy, *Blood Music* is a subtle exploration of the imagination, inspired (but not limited) by ideas loosely related to nanoscience.

Human imagination is far less creative than we often like to think, however, so visionaries and science fiction writers usually imagine any new kind of technology as an extrapolation of older technologies. Where *Prey* and *Blood Music* imagine nanotechnology in terms derived from biology, Neal Stephenson's highly acclaimed futuristic novel, *The Diamond Age*, draws its analogies from computer science. His earlier novel, *Snow Crash*, established Stephenson as a leading writer of cyberpunk fiction—stories that imagine a world transformed by computing, typically written in a pessimistic yet flamboyant style, simultaneously preaching rebellion against advanced technological society while reveling in the technological glories created by it.[14] *The Diamond Age* imagines that the wealthy nanotech barons have created a culture for themselves modeled on that of Victorian England, while the rest of society has fragmented into revived ethnic tribes and cult-like communities. In a premise that actually possesses some merit as a sociological theory, Stephenson explains, "Now nanotechnology had made nearly anything possible, and so the cultural role in deciding what should be done with it had become far more important than imagining what could be done with it."[15]

Given his appreciation of real cutting-edge computer technology, Stephenson is able to write reasonably convincing descriptions of how nanotechnology devices might work, albeit without any formal analysis of whether his poetic notions would really perform as posited. A common home appliance of the future is the microwave-oven-sized "matter compiler," which follows computer-programmed instructions to build up small objects from the nanoscale by feeding precise sequences of microscopic components along networks of tiny conveyors to build the desired object, with hundreds of thousands of nanostructures being created per second. This image is not very different from that of today's "three-dimensional printing" or stereolithography, a rapid prototyping method using complex material feeds and laser beams to sculpt models of industrial products.[16]

Similarly, Stephenson's depiction of "smart paper" imagines a network of simple nanoscale computers woven into a layer between two display surfaces to create something that looks like an ordinary sheet of paper but that can dynamically show any desired text or picture, including animations. Again, while used imaginatively in the story, this idea is not very different from today's paper-like computer displays (sometimes called electronic ink or electronic paper), except for the fact that Stephenson's smart paper does not seem to need a power supply.[17]

Remarkably, the piece of nano-enabled technology that is central to the story is a children's book. One of the neo-Victorian aristocrats notices a profound contradiction in the culture of his mid-twenty-first-century society: Gaining aristocratic status requires a person to be versatile, motivated, courageous, and creative, but the children of aristocrats grow up in secure households that inadvertently teach them to be narrow-minded, passive, timid, and conformist. To overcome this problem, the nanotech baron commissions a team of information and nanotechnology engineers to create *A Young Lady's Illustrated Primer*, an intelligent, interactive book—rather like an instructional quest game—to educate his young daughter. Although only a single copy is supposed to be produced, one of the engineers secretly makes a duplicate for his own daughter. When this book is stolen, it falls into the hands of an impoverished four-year-old girl living in a dysfunctional slum family. Thus the theme of *The Diamond Age* is the possibility that nanotechnology will remake not merely our machines and society, but our very selves.

Although many single science fiction novels about nanotechnology have been published, probably the most significant multivolume work of literature on the topic is the Nanotech Quartet by Kathleen Ann Goonan: I. *Queen City Jazz* (1994); II. *Mississippi Blues* (1997); III. *Crescent City Rhapsody* (2000); and IV. *Light Music* (2002).[18] Although it incorporates many ideas about how "nan" (as Goonan's characters call it) may develop over the coming two centuries, the quartet of novels is highly aesthetic in quality rather than technical in nature. As the titles of the four novels suggest, music is a central theme of the mythos. Characters frequently hear music, and they often sing, play instruments, and dance. One character even has a gun that immobilizes enemies by shooting songs at them. While remaining lucid, the language is often poetic, and many characters briefly experience altered states of perception expressed through stream-of-consciousness writing. Goonan imagines that nano-enabled technology of the twenty-second century will make it possible to create simulacra of deceased individuals who were sufficiently creative to leave behind ample evidence of how their minds worked. Among the characters who participate in important scenes are deceased authors Ernest Hemingway and Mark Twain (two of him arguing with each other!) plus musicians Duke Ellington and Sun Ra. Thus Goonan's quartet is a work of modern surrealist literature, rather than a series of science or technology textbooks. Nonetheless, the Nanotech Quartet suggests ways in which humans might react to the possibilities of nanotechnology, if indeed some of the more radical applications turn out to be feasible.

Goonan says, "Nanotechnology can be enslaving, and it can also be liberating."[19] Also, according to her, with "a viable nanotech . . . humanity would pass a point of no return, beyond which everything would be unimaginably changed."[20] Clearly, Goonan thinks nanotech will be magically powerful, and not merely multiply the effectiveness of conventional technologies. Readers of these novels by Bear, Stephenson, and Goonan do not expect the science-related ideas in the fiction to be true, but rather enjoy the fiction as speculative fantasy. The assumptions about the nanoscale are literary devices that permit the construction of interesting stories, for both fun and aesthetic pleasure.

Our civilization will benefit from artistic creativity that is inspired by nanotechnology. Science fiction can also inspire young people to enter technical fields, and it can communicate to a wide audience the excitement that real scientists and engineers experience in their work. We should applaud the individual writers who have begun promulgating visionary images of the future of the field, and we cannot expect them to limit their imaginations to the proven nanoscale techniques of today. At the same time, these works of fiction are clearly not an appropriate basis for deciding policy, and they may misinform the public about both investment opportunities and potential hazards. The results could be nanocrazes and nanopanics.

DREXLER'S VISION

There is every reason to be ambivalent about enthusiasm in science and technology. On the one hand, without vision, creative people are unlikely to devote the single-minded energy needed to complete great projects. On the other hand, vision can sometimes be an illusion, leading to waste, confusion, and even danger. In terms of the publicity accorded to nanotechnology by mass media and public awareness, K. Eric Drexler has provided the guiding vision. His influence has been tremendous, and a few paragraphs here can hardly do it justice. However, we cannot understand nanotechnology without considering Drexler's conception of it, and an interesting parallel with the history of spaceflight may help us grasp the significance of visionary leaders like him.

Drexler earned his first fame as a prominent member of the L-5 Society. This national organization of young spaceflight enthusiasts was founded in 1975 by Carolyn and Keith Henson, who were inspired by the ideas of Gerard K. O'Neill, a Princeton physicist. O'Neill's vision of space colonization focused on a space city that could be built in the Moon's orbit at a stable point equidistant from the Earth and Moon, trailing the Moon as it orbited the Earth.[21] This was the so-called Lagrange 5 point—hence the name L-5 Society.

As O'Neill and his students developed the idea, the L-5 city was qualitatively quite reasonable but ultimately impractical. Yes, a city built at the L-5 point would stay there indefinitely. Locating it at this spot is a cute idea, the kind that decorates science fiction stories, and probably offers advantages over orbits closer the Earth, because the city was supposed to be built largely of raw materials from the Moon. A mass driver would be built on the lunar surface; this electromagnetic catapult or giant solenoid would use solar power to fling loads of iron or other materials to a mass catcher at L-5.

All of these concepts seem reasonable, if one thinks qualitatively rather than calculating the cost to build the whole system. In 1974, the Skylab project completed successful experiments with an orbiting space station, and in 1975 the last Apollo flight carried out a symbolic rendezvous with a Russian vehicle. Duplicates of Skylab and the Apollo–Soyuz Test Project are on display in the Smithsonian Air and Space Museum, and hundreds of tourists troop through the house-sized Skylab every day. When I completed my doctoral dissertation on the social history of spaceflight in 1975, it seemed reasonable to assume that the forthcoming space shuttle would reduce the cost of launching to Earth orbit quite significantly.[22] In the absence of detailed calculations of the quantity of manufactured materials that would need to be sent from Earth to L-5 and to the lunar surface to build O'Neill's city, in those optimistic days the idea had a ring of plausibility.

In the succeeding three decades, we learned two difficult lessons about spaceflight. First, it turned out to be much tougher to develop a cheap, reliable, reusable space transportation system than many of us had imagined. The space shuttle proved to be the worst kind of failure, the kind that almost succeeds and thereby precludes the development of better alternatives. NASA's half-hearted attempts to develop a second-generation shuttle led nowhere, and the fatal crashes of Challenger and Columbia discredited the very idea of easy spaceflight.

Second, it became clear that the world's ambitions in space are really quite modest. The Apollo program was the result of an international competition for prestige—that is, the space race between the United States and its Cold War opponent, the Soviet Union—and it ended without producing a new rationale for ongoing spaceflight. Neither scientific discovery nor economic benefit had been a significant factor in the Moon effort, and neither motive could be served cost-effectively by an orbiting city. Sometimes one hears the claim that research in the space program will tell us about our origins. In fact, anthropologists have been trying to build a modest program to search for archaeological evidence about human origins in East Africa, where our species first evolved, and their effort could be sustained for a century for the cost of a single space shuttle flight. While 10,000 volunteers might be willing to live at L-5, it is hard to see how their activities while living there could ever hope to pay off the possibly billion-dollar cost of getting each person and his or her possessions there.

In the mid-1970s, memories of the 1973 oil crisis inspired many people to wonder whether Earth's energy needs might be satisfied by solar-powered satellites, perhaps beaming the power down in the form of microwaves to be harvested by antenna farms. These satellites could be built and maintained by crews living at L-5. In addition to the vast costs of launching the original infrastructure required to get a lunar-orbital economy going, many technical problems would inevitably arise in the development of the satellites themselves—not to mention a potentially devastating public relations problem, in that microwave beams from space seem dangerous. Periodically, this idea is resurrected. Indeed, in 2001, I helped manage a joint NASA-NSF research competition to explore the construction of large structures in space, with solar-powered satellites specifically in mind, although we assumed that robots rather than L-5 citizens would do the constructing.

In the October 1976 issue of *L-5 News*, Drexler published a short article, "What Need Not Be Done," listing many things that some people felt would be prerequisites for space colonization but that he thought were unnecessary.[23] The article implied that colonization would be rather easy and that people had made it seem difficult by attaching other goals to it. In the April 1977 issue, in which he was identified as "research assistant to Gerard K. O'Neill," Drexler argued that exploitation of extraterrestrial raw materials could be economic; in conjunction with Keith Henson, he also co-authored an article proposing that vapor deposition methods similar to those employed in making integrated circuits could be used to construct large structures in the vacuum of space.[24] The following year Drexler published two articles, one about the legal and ethical aspects of exploiting extraterrestrial resources, and a second suggesting how to respond to questions about the safety of solar satellites' microwaves.[25]

I remember being invited to a meeting of the MIT L-5 group to give a talk about my research on the spaceflight social movement, of which L-5 was a latter-day part. My main thesis was that conventional motives were unlikely to take humanity into space—neither scientific nor economic goals, and not widespread popular enthusiasm. Members of the spaceflight social movement achieved the successes they did by manipulating political leaders, presenting their rocket ideas as if they were solutions to problems the leaders faced. *The Spaceflight Revolution*, my doctoral dissertation published in 1976, was a study of the social movement that achieved spaceflight.[26] In the issue of *Time* magazine marking the tenth anniversary of the first Moon landing, Arthur C. Clarke summarized my thesis as follows:

> As William Sims Bainbridge pointed out in his 1976 book *The Spaceflight Revolution: A Sociological Study*, space travel is a technological mutation that should not really have arrived until the 21st century. But thanks to the ambition and genius of Wernher von Braun and Sergei Korolyev, and their influence upon individuals as diverse as Kennedy and Khrushchev, the moon . . . was reached half a century ahead of time.[27]

Actually, my thesis was that the Moon might never have been reached, because the period 1945–1965 was a launch window during which the development of the intercontinental nuclear missile could advance the technology needed for spaceflight. Had humanity been wise enough never to develop the atomic bomb, the expensive ICBM would never have made sense as a weapon. Had leaders of the spaceflight movement delayed their push for even a decade, the cruise missile would have supplanted the ICBM, and warheads would have become light enough that the first generation of very large missiles would never have been necessary. Those rockets were the ones that put the first men in orbit, and their emergence led to the large liquid-fueled rocket engines that are the best options for launching spacecraft.

In his 1951 science fact book *The Exploration of Space*, Clarke had suggested that spaceflight would begin gradually some time in the twenty-first century, when the Earth's technology and economy had developed to the point that this feat could be achieved without excessively heroic efforts.[28] Many people believe that when a technology becomes possible, it will inevitably be developed. My book specifically argued against this form of technological determinism. Scientific and technological revolutionaries must often work around the inertia of other people in society, selling their ideas in somewhat distorted terms to get needed support. Wernher von Braun (in both Nazi Germany and the United States) and Sergei Korolyev (in the Soviet Union) were both members of amateur spaceflight clubs and were dedicated to this lofty dream for personal and emotional reasons. No one was willing to pay to have their spaceships built, so von Braun and Korolyev repackaged them as weapons and sold them to leaders like Hitler, Stalin, Kennedy, and Khrushchev at times when these powerful leaders needed a way to overcome political adversity. Whenever progress in the space program is blocked, visionaries need to find a detour around public indifference, such as the loss of interest after the Apollo missions.

The first space shuttle orbital flight took place in 1981. Well before the Challenger disaster in January 1986, it had become obvious that space technology would develop rather more slowly than L-5 members wanted. Blocked in their attempt to achieve exciting careers in the conquest of space, many of them began exploring other areas. Drexler came across a remarkable lecture by the famous physicist Richard Feynman, "There's Lots of Room at the Bottom," meant to inspire young scientists about the opportunities for accomplishment. In 1981, Drexler published a journal article, the very first word of which is "Feynman," arguing that proteins and other microbiological structures could be treated as components from which engineers could build things.

Drexler's article uses the term "microtechnology" rather than "nanotechnology" to describe this building process. Among its radical assertions are three paragraphs implying that molecular devices could undo the harm caused by freezing the human body, presumably referring without citation to the cryonics movement's idea of freezing people so that they might be revived in the future when technology has advanced enough to cure the person's diseases. This is another science fiction idea found in such classic stories as "The Jameson Satellite" by Neil R. Jones and "The Resurrection of Jimber-Jaw" by Edgar Rice Burroughs,[29] and in movies such as *Buck Rogers* (1939) and *The Thing* (1951). The visionary who did the most to develop the cryonics ideology was Robert C. W. Ettinger, in his books *The Prospect of Immortality* (1964) and *Man into Superman* (1972), as well as the article "Interstellar Travel and Eternal Life," published in a 1968 issue of the science fiction magazine *If*.[30]

Drexler eventually expanded further on Feynman's ideas in his visionary book *Engines of Creation: The Coming Era of Nanotechnology*, which was published in 1986. This work was not a technical monograph, and it did not report the results of any research. Rather, it was a trade book, intended for a wide audience.[31]

Drexler's book does not cite Taniguchi's 1974 paper, and it is possible he coined the word "nanotechnology" independently, but he defines it rather differently from Taniguchi. By the mid-1980s, as explained more fully in Chapter 3, the components on integrated circuit chips had been shrinking rapidly for a quarter of a century, justifying the generic term "microelectronics" and the widespread prediction that they would continue to get smaller and smaller. This is the context in which Drexler introduced the word "nanotechnology":

> *Micro*circuits have parts measured in *micro*meters—that is, in millionths of a meter—but molecules are measured in *nano*meters (a thousand times smaller). We can use the terms "nanotechnology" and "molecular technology" interchangeably to describe the new style of technology. The engineers of the new technology will build both nanocircuits and nanomachines.[32]

In the same year his first book was published, Drexler founded the Foresight Institute to promote nanotechnology. That institute still exists today, and its website proclaims that it was "the first organization to educate society about the benefits and risks of nanotechnology."[33] By 1991, Drexler had become sufficiently well known that the journal *Science* published an article about him, calling him both "the apostle of nanotechnology" and a "flake" of "nanoreligion."[34]

The title of Drexler's second book, which was published five years after *Engines of Creation*, expresses vast hope: *Unbounding the Future: The Nanotechnology Revolution*. Like its predecessor, this work was a trade book aimed at a wide audience, and it did not make a careful case for the feasibility of its ideas based on laboratory research. This time, Drexler had a co-author, Christine Peterson, who had a college degree in chemistry. At the heart of the vision was a nanoscale machine capable of doing any kind of work:

> The idea of nanotechnology begins with the idea of a *molecular assembler,* a device resembling an industrial robot arm but built on a microscopic scale. A general-purpose molecular assembler will be a jointed mechanism built from rigid molecular parts, driven by motors, controlled by computers, and able to grasp and apply molecular-scale tools. Molecular assemblers can be used to build other molecular machines—they can even build more molecular assemblers. Assemblers and other machines in molecular-manufacturing systems will be able to make almost anything, if given the right raw materials.[35]

Unbounding the Future is dominated by this image of nanotechnology: nanoscale robot arms assembling nanomachines by fitting together individual atoms as if they were Lego pieces. However, Drexler periodically says that other methods of nanoscale engineering may be available—for example, techniques based on the principles of biochemistry. He expresses great passion for his "unbounded" vision of the potential of nanotechnology, but he has sufficient perspective to realize that his own technical approach might prove infeasible. In short, Drexler urges the reader to accept the general vision quite apart from the specific means to achieve it.

Drexler's third book, *Nanosystems,* was far more technical, purporting to offer a comprehensive explanation of how the author's vision of nanotechnology could be realized. The publisher was Wiley-Interscience, which I must believe is a reputable scientific publisher because coincidentally it published my own first book. The copy of *Nanosystems* that I read belonged to the Pentagon library, suggesting it was a scientific reference for the U.S. military establishment. The preface explains Drexler's approach:

> Molecular manufacturing applies the principles of mechanical engineering to chemistry (or should one say the principles of chemistry to mechanical engineering?) and uses results drawn from materials science, computer science, and elsewhere.[36]

This is a very important statement of Drexler's perspective for two reasons. First, it is a classic expression of converging technologies—that is, taking ideas from one field and applying them to another, thereby binding the two fields together. Second, it precisely defines Drexler's revolutionary concept:

Nanotechnology is chemistry performed using the principles of mechanical engineering. The book contains many diagrams of hypothetical mechanical parts such as bearings, gears, and even screws that would be composed of small numbers of atoms. Unlike Drexler's two earlier books, *Nanosystems* cites many scientific publications, although many sections and specific scientific claims lack supporting references. This is an earnest book, fully 500 large pages that prove the author was very serious in his work and honestly hopeful that his approach could be technically successful.

According to Drexler's online Wikipedia biography (which presumably he has had the opportunity to correct if it contains substantive errors), *Nanosystems* was an outgrowth of its author's Ph.D. work at MIT. The biography says Drexler earned two doctorates in 1991, one in architecture and the other "on the topic of molecular nanotechnology" from the MIT Media Lab. On the basis of my own visits, I can say that the Media Lab is a marvelous place that encourages creativity in science and engineering, by both students and faculty. The projects I have seen there all dealt with computer-related areas, especially new kinds of human–computer interaction, rather than with nanotechnology. The Media Lab currently emphasizes robot work, and it may have done so in Drexler's day as well. It is definitely not a chemistry lab, and Drexler's one technical book does not report the results of experiments actually working at the molecular scale. An architecture degree suggests that its owner has the ability to imagine and describe physical structures on the large scale, involving many of the skills a mechanical engineer would possess, but it also has nothing to do with nanoscale chemistry. This is not to say that Drexler is incompetent at the nanoscale, but is merely meant to emphasize that he brings a very different kind of expertise to his theorizing than that possessed by physicists, chemists, and materials scientists.

In 2004, *The Economist* called Drexler the "father" and "inventor" of nanotechnology, reporting he was unhappy that the name of his idea had been stolen and applied to something else—namely, ordinary science and technology carried out at the nanoscale.[37] A glossary at the end of *Nanosystems* recognized already a dozen years earlier that Drexler had been unable to impose his vision of nanotechnology upon professional scientists in the most closely related fields, by defining "nanotechnology" twice:

> In recent general usage, any technology related to features of nanometer scale: thin films, fine particles, chemical synthesis, advanced microlithography, and so forth. As introduced by the author, a technology based on the ability to build structures to complex, atomic specifications by means of mechanosynthesis; this can be termed molecular nanotechnology.[38]

ALCHEMY AT THE NANOSCALE

If assemblers can make anything, then they can make more assemblers and they can make the machinery needed to get raw materials for the assemblers to use in assembling. Thus an assembler factory can become self-sustaining, as if it were a new form of life. So long as it continues to follow human instructions, it can then make all the products humans want.

Modern societies have been called postindustrial, in recognition of the fact that manufacturing has become a smaller part of the economy than services, including medical and information services.[39] Given this transformation of society, one might think that the economic benefit of nanotechnology would be limited to reducing the cost of manufactured goods to zero, leaving most of the economy untouched. In the Drexlerian vision, however, assemblers can build computer-controlled nanoscale medical robots that then kill cancer cells, clean out the circulatory system, and perform dozens of other life-saving medical tasks. In this way, nanotechnology might replace some industries currently considered services and return manufacturing to the center of the economy. John Horgan is scornful of such notions:

> As espoused by evangelists such as Eric Drexler, nanotechnology resembles a religion more than a field of science. Drexler and others proclaim that we will soon be able to reconstruct reality from the atomic scale on up in ways limited only by our imaginations. Our alchemical power to transform matter will help us achieve infinite wealth and immortality, among other perks.[40]

As mentioned in Chapter 1, Horgan was a technological pessimist who believed that science had already elucidated much of what there was to discover. The same cannot be said of Richard Smalley. The Wayback Machine of the Internet Archive allows us to visit webpages of the past, and the earliest page for Drexler's Foresight Institute (dating from November 5, 1996) announces Smalley's Nobel Prize for helping to discover fullerenes in 1985.[41] Perhaps ironically, Smalley became Drexler's most effective critic when both contributed essays to a special issue of *Scientific American* in 2001.

Drexler connected nanotechnology to his earlier dream of space colonization: "Low-cost, lightweight, extremely strong materials would make transportation far more energy efficient and make space transportation economical. The old dreams of expanding the biosphere beyond our one vulnerable planet suddenly look feasible once more." This is a good point, and nanotechnologists of every stripe would be ready to say they think it is technically possible. Drexler then enters far more controversial territory: "Medical nanorobots are envisioned that could destroy viruses and cancer cells, repair damaged structures, remove accumulated wastes from the brain, and bring the body back to a state of youthful health." He also says that nanotechnology might be used to bring back to life the legally dead people who had their bodies frozen in cryonic suspension, promotes his idea of nanoscale assemblers that could manufacture anything including themselves, and warns of the need to avoid abuse of this potent technology.[42]

Smalley's essay, which was based on his contribution to our original report on the societal implications of nanotechnology, argues that Drexler's nanoscale robots are physically impossible, listing a very large number of flaws in the idea. He calculates that even if these robots operated at high speed, "generating even a tiny amount of a product would take a solitary nanobot millions of years," because individual atoms and molecules are so small. Thus it would be necessary to work with large numbers of assemblers simultaneously, which raises thorny issues related to supplying those robots with the needed raw materials (one atom at a time) and controlling their behavior.

Smalley then raises two issues he says are fundamental to the feasibility of nanoscale robots: the facts that the finger of an assembler's arm will be too fat and too sticky. The fingers are fat because it takes many atoms to build a structure complex enough to do the work, up in the scale of several nanometers, but the individual atoms they are supposed to manipulate are much smaller than a nanometer. Thus the geometry of the assembler does not work. It is as if the automatic pin-setting machine in a bowling alley must be made entirely out of bowling balls. The sticky finger problem arises because "the atoms of the manipulator hands will adhere to the atom that is being moved. So it will often be impossible to release this miniscule building block in precisely the right spot."[43]

In 2003, Drexler and Smalley moved their debate to the pages of *Chemical and Engineering News*.[44] Drexler accused Smalley of misrepresenting his concept of nanotechnology, though I believe Smalley was innocent of that crime. Appropriately, Smalley focused on the mechanical engineering model of molecular nanotechnology, including the assembler that dominated Drexler's books and had the greatest impact on the mass media and science fiction. The fact that Drexler also discussed a range of other approaches—notably, modeling nanotechnology on molecular biology—means that he has some justification for complaining that Smalley should not dismiss the idea altogether. In this book, we will consider the biological approach in Chapter 4.

In response, Smalley first apologized if he had misrepresented Drexler's ideas, confessing that he had been inspired by Drexler's first book when he read it in 1991. He then offered several criticisms of a biological approach based on enzymes or ribosomes. In particular, he noted that Drexler's books represented nanotechnology as happening in a dry environment, like a conventional assembly-line factory, when enzymes and ribosomes actually need to work in water. Further, Smalley noted, biological systems emphasize relatively few chemical elements and have great difficulty working with silicon, copper, aluminum, titanium, and many other elements needed in industry. This exchange between Drexler and Smalley left them farther apart than ever.

The professional nanotechnologists I have talked with agree with Smalley's position and reject Drexler's view. Nevertheless, it remains true that Drexler may have inspired both scientific nanotechnology research and investment in the field. Writing in the journal *History and Technology*, W. Patrick McCray suggests that Drexler was the key visionary who led to the emergence of a nanotechnology movement, and posits that his testimony to a congressional hearing organized by then-Senator Albert Gore was influential in setting the new technologies agenda of the U.S. government.[45]

Political scientists should examine the origins of the National Nanotechnology Initiative to see if one source of support among policy makers and legislators was Drexler's vision, even though the NNI's billion dollars a year is not flowing into Drexlerian projects. When I recently searched the website of the NNI for the word "Drexler," his name did not appear on the site. Similarly, Drexler is not mentioned in one of the most authoritative popular books on the subject, *Nanotechnology: A Gentle Introduction to the Next Big Idea* by Mark and Daniel Ratner.[46] Some of Drexler's supporters expressed irritation that Smalley did not examine their hero's ideas more deeply, but unlike those of us who worked with him, they apparently did not know that Smalley was dying of leukemia and had more pressing issues on his mind. Smalley passed away October 28, 2005, before Drexler's nanoscale robots could be developed to cure him.

NANOPANIC

If nanotechnology can be an apt arena for a craze, it can also be a fertile battleground for a panic. The word "panic" describes situations in which terrified people attempt to flee danger. Crazes and panics are mirror images of each other: In a craze, a group of people rushes *toward* something they all desire; in a panic, they rush *away* from something they all fear. Panic can be inspired by a real danger, particularly one that is immediate and difficult to avoid (such as a fire in a crowded theater), but it can also be caused by imaginary dangers and rumors of danger. One thinks of the Martian invasion panic triggered by the 1938 radio dramatization of H. G. Wells' science fiction story, *The War of the Worlds*, produced by actor Orson Welles and his Mercury Theatre company, telling the story as if it were really happening. However, there is good reason to believe that the alleged panic was an invention of the news media. In 1973, a Swedish radio dramatization of the explosion of a nuclear power plant supposedly caused mass panic, but careful social-science research discovered that no panic occurred outside the mass media.[47] Clearly, reports of nanopanic must be taken with a nanoscale grain of salt.

In an article in *Wired* magazine published right after April Fool's Day 2000, but writing in deadly seriousness, computer software designer Bill Joy warned that nanotechnology—combined with genetic engineering and robotics—endangered human survival. In part he was responding to Ray Kurzweil's prophecy that humans would merge with their computers within the next few years to initiate the Age of Spiritual Machines.[48] Joy assumed it would soon be possible to build self-reproducing nanoscale robots, accepting the "gray goo scenario" that nanobots would destroy all living things. His article received wide publicity and immediately became available over Internet. Similar hysterical ideas were disseminated by other writers, continuing through the publication of Michael Crichton's novel *Prey* two years later. We can wonder if a widespread nanopanic ensued.

At the end of 2001, a team of sociologists led by James Witte of Clemson University administered a major international questionnaire over Internet. As a member of the team, I contributed a set of questions about science and technology, including an agree–disagree item based on language in Joy's article: "Our most powerful 21st-century technologies—robotics, genetic engineering, and nanotechnology—are threatening to make humans an endangered species." Only 9.0 percent of 3,909 respondents agreed with this statement, and among respondents who had graduated from college only 5.6 percent agreed. The questionnaire project had been sponsored by the National Geographic Society, and respondents were better educated than average, tended to be relatively knowledgeable about technology-related issues, and included substantial numbers of people who were concerned about protecting the natural environment. Thus the respondents were not a random sample, so we cannot estimate exactly what fraction of the total population is worried about human extinction from nanotechnology. But given that the Bill Joy item quotation cited hazards of genetic engineering, about which reasonable and well-informed people may legitimately have concerns, and mentioned robots, then nanopanic must be (appropriately) miniscule.[49]

The National Geographic online questionnaire also asked a subset of respondents to write their thoughts freely about both this item and another agree–disagree item, "Human beings will benefit greatly from nanotechnology, which works at the molecular level atom by atom to build new structures, materials, and machines." Many respondents described specific benefits they thought might come from nanotechnology, and a few mentioned problems that might arise. One of their chief concerns was the reasonable possibility that nanotechnology, if applied inappropriately, might further erode our privacy. A few mentioned the issue of whether the benefits of nano would be widely shared or would contribute to deleterious inequality between classes. One or two wrote vaguely of unknown dangers or wondered whether humans possess the maturity to use nanotechnology well. Not a single one of the 598 respondents who wrote English-language comments about nanotechnology mentioned gray goo or any other specific way in which nanotechnology might threaten human existence. Apparently, to this point real nanopanic does not exist, even if a few professional fear-mongers would like us to think otherwise.

CONCLUSION

One theme of this book was firmly established in Chapter 1: In our still-so-primitive culture, it is difficult to marshal the energies need to sustain scientific progress. In this chapter we have established a related theme: Nanotechnology can gain support for science more broadly by attracting people's uninhibited hopes. Call it fantasy, if you will, but fantasy is one of humanity's greatest sources of real innovation.

I share with Drexler a profound dissatisfaction that our society seems incapable of exploring and exploiting the solar system at anything like the pace permitted by existing science, let alone the pace that might be achieved by a fully mature nanoscience. To sell the idea of the space program, visionaries of the past were forced to "oversell" its benefits. If someone refuses to listen, you must shout to gain his or her attention. Hence space pioneers like Wernher von Braun proclaimed that spaceflight would be easier than it turned out to be, and science fiction writers told exciting stories that raised unrealistic hopes about the human goals that could be achieved on other planets. This overselling was not the same as lying, however. People of the twentieth century simply did not have the context of knowledge or imagination to appreciate the value of new things that could be achieved in outer space, so we had to explain their value in metaphors like the silly notion of mining the asteroids for their iron. The most valuable resources of outer space do not even have names yet. A debate between the future and the past is bound to contain many anachronisms.

Nanotechnology may, in fact, make space travel easier, and it was a reasonable course for people like Drexler (and, quite frankly, myself) to take a nanotech detour on the way to the stars, aiming for the infinite by way of the infinitesimal. Drexler's universal nanoscale assemblers are a metaphor comparable to the L-5 city of his mentor, Gerard K. O'Neill. Qualitatively, we can conceive of each of them, and they stretch our thinking in new directions. Nevertheless, the necessary scientific and economic details to judge their practicality remain lacking. Nanotechnology will accomplish many wonderful things in novel manners. The assemblers are an imaginative placeholder for the realities to come. There is no telling which of the wildest ideas of science fiction might turn out to be surprisingly feasible, or which ideas will inspire young people to become scientists and engineers, building the future of humanity.

REFERENCES

1. Tom Godwin, *The Cold Equations and Other Stories* (New York: Baen, 2003).

2. Steven Weinberg, *The Discovery of Subatomic Particles* (New York: W. H. Freeman, 1990).

3. James Blish, *The Seedling Stars* (New York: Gnome, 1957).

4. Isaac Asimov, *Fantastic Voyage* (New York: Bantam, 1966). The original story is attributed to a lesser-known science fiction writer, Jerome Bixby, and was developed further by screen writers before Asimov novelized it from the 1966 Twentieth Century Fox film.

5. Isaac Asimov, *I, Robot* (New York: Grosset and Dunlap, 1950, p. 7).

6. Daniel C. Dennett, *Darwin's Dangerous Idea* (New York: Simon and Schuster, 1995, p. 500).

7. William Sims Bainbridge, *Dimensions of Science Fiction* (Cambridge, MA: Harvard University Press, 1986, p. 42), based on hearing Asimov tell the story.

8. Arthur C. Clarke, "Hazards of Prophecy," in Alvin Toffler (ed.), *The Futurists* (New York: Random House, 1972, p. 144); cf. Arthur C. Clarke, *Profiles of the Future* (New York: Bantam, 1963).

9. Sam Moskowitz, *Explorers of the Infinite* (Cleveland: Meridian, 1963, p. 11).

10. Michael Crichton, *Jurassic Park* (New York: Knopf, 1990).

11. Michael Crichton, *Prey* (New York: Harper Collins, 2002, p. x).

12. William Sims Bainbridge, *Dimensions of Science Fiction* (Cambridge, MA: Harvard University Press, 1986, p. 24).

13. Greg Bear, "Blood Music," *Analog*, 106:12–36, 1983; *Blood Music* (New York: Simon and Schuster, 2002).

14. Neal Stephenson, *Snow Crash* (New York: Bantam Books, 1992).

15. Neal Stephenson, *The Diamond Age, or, Young Lady's Illustrated Primer* (New York: Bantam Books, 1995, p. 37).

16. K. G. Cooper, *Rapid Prototyping Technology: Selection and Application* (New York: Marcel Dekker, 2001); C. K. Chua, K. F. Leong, and C. S. Lim, *Rapid Prototyping: Principles and Applications* (New Jersey: World Scientific, 2003); P. K. Venuvinod and W. Ma, *Rapid Prototyping: Laser-Based and Other Technologies* (Dordrecht, Netherlands: Kluwer, 2004).

17. Gregory Philip Crawford, "Electronic Paper Technology," in William Sims Bainbridge (ed.), *Encyclopedia of Human–Computer Interaction* (Great Barrington, MA: Berkshire, 2004, pp. 205–208).

18. Kathleen Ann Goonan, *Queen City Jazz* (New York: Tor, 1994); *Mississippi Blues* (New York: Tor, 1997), *Crescent City Rhapsody* (New York: Eos, 2000); *Light Music* (New York: Eos, 2002).

19. Kathleen Ann Goonan, *Mississippi Blues* (New York: Tor, 1997, p. 511).

20. Kathleen Ann Goonan, *Crescent City Rhapsody* (New York: Eos, 2000, p. 6).

21. Gerard K. O'Neill, "The Colonization of Space," *Physics Today*, (September): 32–40, 1974; *The High Frontier: Human Colonies in Space* (New York: Morrow, 1977).

22. William Sims Bainbridge, *The Spaceflight Revolution* (New York: Wiley Interscience, 1976).

23. "What Need Not Be Done," *L-5 News*, 14:7–8, 1976.

24. Eric Drexler, "Non-terrestrial Resources," *L-5 News*, 2(3):4–5, 1977; Keith Henson and Eric Drexler, "Vapor Phase Fabrication of Structures in Space," *L-5 News*, 2(3):6–7, 1977.

25. K. Eric Drexler, "Space Mines, Space Law, and the Third World," *L-5 News*, 3(4):7–8, 1978; "Microwaves: SPS Hazard," *L-5 News*, 3(5):5–6, 1978.

26. William Sims Bainbridge, *The Spaceflight Revolution* (New York: Wiley Interscience, 1976).

27. Arthur C. Clarke, "The Best Is Yet to Come," *Time*, July 16, 1979, p. 27.

28. Arthur C. Clarke, *The Exploration of Space* (New York: Harper, 1951).

29. Neil R. Jones, "The Jameson Satellite," *Amazing Stories*, 30:156–176, 1956, reprinted from the July 1931 issue; Edgar Rice Burroughs, "The Resurrection of Jimber-Jaw," *Argosy*, February 20, 1937.

30. Robert C. W. Ettinger, *The Prospect of Immortality* (Garden City, NY: Doubleday, 1964); *Man into Superman: The Startling Potential of Human Evolution—and How to Be Part of It* (New York: St. Martin's Press, 1972); "Interstellar Travel and Eternal Life," *If*, 18:109–114, 1968.

31. K. Eric Drexler, *Engines of Creation* (Garden City, NY: Anchor Press/Doubleday, 1986).

32. K. Eric Drexler, *Engines of Creation* (Garden City, NY: Anchor Press/Doubleday, 1986), http://www.e-drexler.com/d/06/00/EOC/EOC_Chapter_1.html

33. http://www.foresight.org/; retrieved July 15, 2006.

34. Ivan Amato, "The Apostle of Nanotechnology," *Science*, 254:1310–1311, 1991.

35. K. Eric Drexler and Chris Peterson, *Unbounding the Future: The Nanotechnology Revolution* (New York: William Morrow, 1991, p. 33).

36. K. Eric Drexler, *Nanosystems: Molecular Machinery, Manufacturing, and Computation* (New York: Wiley-Interscience, 1992, p. xviii).

37. "Nanotechnology's Unhappy Father," *The Economist*, March 13, 2004, pp. 41–42.

38. K. Eric Drexler, *Nanosystems: Molecular Machinery, Manufacturing, and Computation* (New York: Wiley-Interscience, 1992, p. 527).

39. Daniel Bell, *The Coming of Post-Industrial Society* (New York, Basic Books, 1973).

40. John Horgan, "The End of Science Revisited," *Computer*, 2004, 37(1), p. 38.

41. http://web.archive.org/web/19961105125543/http://www.foresight.org/

42. K. Eric Drexler, "Machine-Phase Nanotechnology," *Scientific American*, 285(3):74–75, 2001.

43. Richard E. Smalley, "Of Chemistry, Love and Nanobots," *Scientific American*, 285(3):76–77, 2001; "Nanotechnology, Education, and the Fear of Nanobots," in Mihail C. Roco and William Sims Bainbridge (eds.), *Societal*

Implications of Nanoscience and Nanotechnology (Dordrecht, Netherlands: Kluwer, 2001, pp. 145–146).

44. "Nanotechnology: Drexler and Smalley Make the Case for and against 'Molecular Machines," *Chemical and Engineering News*, 81(48):37–42, 2003.

45. W. Patrick McCray, "Will Small Be Beautiful? Making Policies for Our Nanotech Future," *History and Technology*, 21(2):177–203, 2005.

46. Mark Ratner and Daniel Ratner, *Nanotechnology: A Gentle Introduction to the Next Big Idea* (Upper Saddle River, NJ: Prentice Hall, 2003).

47. William Sims Bainbridge, "Collective Behavior and Social Movements," in Rodney Stark (ed.), *Sociology* (Belmont, CA: Wadsworth, 1985, pp. 492–523).

48. Ray Kurzweil, *The Age of Spiritual Machines: When Computers Exceed Human Intelligence* (New York: Viking, 1999).

49. William Sims Bainbridge, "Public Attitudes toward Nanotechnology," *Journal of Nanoparticle Research*, 4:561–570, 2002.

Chapter 3

Information Technology

Given adequate investments in research and development, nanotechnology could allow the power of computers to continue to double every two years for another two decades. The future of computing depends upon Moore's law, a prediction about increasing integrated circuit performance that was originally proposed in 1965 by Gordon Moore. At the time, Moore was director of the research and development laboratory of Fairchild Semiconductor, one of the companies that was leading the revolution in electronics. Soon afterward, he helped found Intel Corporation, which has long dominated computer chip manufacture.

MOORE'S LAW

It may be difficult for young people today to envision what electronics was like in the middle of the twentieth century. Radios, televisions, record players, and similar devices contained physically large circuits of a few separate components, of which the flagships were vacuum tubes. Thomas Edison himself had discovered that a light bulb filament, when heated red hot in a near-vacuum, would give off a stream of electrons. Later researchers had figured out that the flow of these electrons could be controlled by changing the electric charge on a grid placed between the filament (called the cathode) and a collector plate (called the anode). A signal sent to the grid could be amplified by the tube, and a number of tubes could be connected through other components to create the radio or other electronic device. Inside its wooden or plastic case, a radio comprised a metal chassis with perhaps half a dozen sockets holding tubes, plus an intricate network of heavy wires connecting resistors, capacitors, coils, and other components.

Traditionally, electronic equipment was assembled by hand. A resistor, for example, was a little component, often the size of an inch-long section of soda straw, with a wire protruding from each end. Depending on the circuit, the worker might tightly wrap the end of one wire around a tab sticking out of the bottom of the socket holding one vacuum tube, using a soldering gun to melt a little drop of metal to hold the wire in place. The same worker would then attach the wire at the other end of the resistor to a tab on a different tube socket using the same method. Tens of thousands of people made electronic equipment as a hobby during this period, including me. In my teens I soldered together a stereo system and several pieces of test equipment, including high-frequency oscillators.

When transistors were first introduced in the 1950s, the construction method was the same: I can remember soldering together a Moog four-transistor theramin kit, an electronic musical instrument that made weird violin-like tones when the player waved his or her hands near it.

Transistors did not have the filaments that made the vacuum tubes run hot; they required much less power, and they were much smaller. By the end of the 1950s, Jack Kilby at Texas Instruments and others had figured out how to manufacture circuits as integrated units, including several transistors on one piece of material—the proverbial "chip."

The standard manufacturing method for integrated circuits no longer required laborious manual assembly with a soldering gun, but rather relied on photolithography. The printing industry had used a similar process, called photoengraving, to produce metal printing plates from photographs since the late nineteenth century. In this technique, a metal plate is covered by a light-sensitive material called the photoresist. A photograph is projected onto it briefly, and the photoresist hardens where the light touches it. Unhardened photoresist is then washed away, and acid is applied to etch the exposed parts of the plate. Next, the remaining photoresist is removed. Ink is picked up by the etched portions of the plate and then applied to paper in the printing process.

Before integrated circuits were developed, the electronics industry had begun using this method to make printed circuits that replaced the tangle of wires in electronic devices; unfortunately, this technique still required soldering on the components. By contrast, today's photolithography comprises a highly developed set of processes that create integrated circuits of great complexity at low cost, with all the components being built in.

Given its origins in the printing industry, photolithography suggests a valid way of conceptualizing modern chip making. It is a form of printing, whose development was just as revolutionary as the invention of the printing press. It has allowed huge improvements in circuit performance with a vast reduction in cost and greatly improved reliability. As a 2005 press briefing from Intel says, "[T]he price of a transistor is now about the same as that of one printed newspaper character."[1] Integrated circuits practically ended the heroic era of amateur construction of electronics. Teenage boys are no longer proud to have solder on their shoes. Integrated circuits also devastated the electronic repair business, where much of the work consisted of replacing burned-out tubes, resistors, and capacitors. In 1974, I purchased a Hewlett-Packard electronic calculator for $400, which in today's money would be something like $1,500. I could buy a replacement today for about 1 percent of the original cost—except that the old machine still works perfectly.

Moore's law concerns the rate of improvement of integrated circuits produced by photolithography. Its original statement came in a 1965 magazine article titled "Cramming More Components onto Integrated Circuits," in which Moore considered a number of factors that would influence progress over the following years, and in which he predicted that one benefit would be the development of home computers. He first phrased the law in this way: "The complexity for minimum component costs has increased at a rate of roughly a factor of two per year."[2] A decade later, when home computers were about to become a reality, Moore reported, "Complexity of integrated circuits has approximately doubled every year since their introduction."[3] During this period, chips were getting larger while individual components were getting smaller. Today, Intel's definition of Moore's law states that the number of transistors on a chip doubles roughly every two years, but other definitions talk about 18 months rather than one or two years, transistor density rather than number of transistors, and doubling of the speed of information processing for a given cost.

I have often heard computer scientists refer to Moore's law in reverential tones, as if it were one of the Ten Commandments: We have a duty to keep integrated circuit complexity doubling every two years. Indeed, a major driver of human history since 1965 has been the rapidly increasing performance of computers and related technologies. At the risk of oversimplifying, we can note the major computer applications of recent decades. In the 1970s, really extensive use of big computers swept industry and government. This era was followed by the introduction of personal computing in the home and school during the 1980s. The World Wide Web brought Internet to everyone in the 1990s, and the first decade of the new century has been marked by mobile and ubiquitous computing. Indeed, old media are converging, as digital technology erases the differences between a camera and a telephone, a computer and a television set, Internet and the radio spectrum.

A key factor in sustaining Moore's law has been the ability to use photolithography to make increasingly smaller components. Intel's 4004 chip, produced in 1971, contained 2,300 transistors; by comparison, Intel's Itanium 2 chip, produced in 2004, contained 592 million transistors.[4] In 1965, Moore noted that it was then possible to build a circuit in which the distance from the center of one transistor to the center of the next transistor was 2/1,000 inch, or about 0.05 millimeter (= 50 microns = 50,000 nanometers). By 2006, however, Intel was able to produce a memory chip containing more than 1 billion transistors, fitting each group of six transistors into a space of 0.346 square micron. The smallest manufactured feature in one of these transistors is only 45 nanometers across.[5] Thus Moore's law has taken microelectronics down into the nanoscale.

At present, Moore's law is approaching its natural limits. Indeed, it is surprising that photolithography has been able to produce such small transistors reliably. Visible light has a wavelength between 700 nanometers (red) and 400 nanometers (violet). In the past, some had suggested that we could not focus light with precision nearly as tightly as the wavelength. In fact, the implementation of a number of optical tricks, plus using ultraviolet light with a wavelength of less than 200 nanometers, has proven remarkably successful in this regard. Experiments are exploring the potential of shorter-wavelength X-rays, and some observers predict that transistors with features as small as 15 nanometers may be possible. However, far down into the nanoscale, we begin dealing with small numbers of atoms, and it may not be possible to sustain the accustomed high reliability of components. In consequence, scientists are exploring many alternative forms of nanoelectronics that may not use photolithography—for example, using carbon nanotubes as transistors and using some kind of self-assembly technique to arrange those nanotubes in circuits.

Another serious problem related to the increasingly small size of integrated circuits is the potential for overheating. Computation takes energy, which ultimately is dissipated as heat. For a long time, this issue has posed a challenge for high-performance computers, and most personal computers today have fans to force air cooling of the central processing chip. Everyone is familiar with the challenge of cooling automobile engines: Except for some low-power engines, air cooling is not sufficient to prevent overheating in these engines. Thus we put water in the radiators of our cars to prevent a "meltdown."

Some large computers use comparable cooling methods. Years ago, I was shown a Cray-2 supercomputer created by the leader in that field, Seymour Cray. This monster used a cooling fluorocarbon liquid called *fluorinert*, which was routed out of the computer and into a rather dramatic glass-enclosed "waterfall" that stood apart from the rest of the machine. I asked the technician showing me the Cray-2 whether it was technically necessary to make the cooling liquid visible in this way or whether its placement was just a public relations gesture. He agreed, jokingly, "That's so you can see more Cray."

What made cooling such a challenge for the Cray-2 was that Cray was trying to pack circuit boards close together, so that they could interact more quickly. Unfortunately, that structure blocked the circulation of cooling air. Liquid under pressure was Cray's answer to this problem.

Today's supercomputers consist of many separated units that divide the calculation among them, doing much work in parallel but without the need for constant, direct communication. This scheme requires the computer program to separate the calculations into many subunits, which can then be carried out simultaneously by separate processing units. While this strategy works for some problems in science and engineering, parallel processing remains very inefficient for many other kinds of problems, at least as we know how to handle them today. Many conceptions of how we might achieve great computation speed with nanoelectronics or molecular computing call for us to escape the two-dimensional confines of silicon chips and perform computations in three-dimensional solids. Unless engineers can find very clever countermeasures, however, this approach would aggravate the problem of overheating.

Quite apart from the significant scientific and technical challenges associated with developing nanotube-based electronics or some other form of molecular computing, there may exist a daunting economic barrier. If the new technology is as complex as the old technology, a huge investment could be required to switch from microelectronics to nanoelectronics. In June 2006, Intel opened a new factory to produce chips with the smallest current mass-production resolution (64 nanometers) on the largest currently produced wafers (300 millimeters). Sited in Ireland to reduce costs, the factory required an investment of $2 billion.[6] Almost simultaneously, the firm's competitor AMD announced an investment of $5.2 billion to set up a factory in New York to produce chips with a claimed 32-nanometer resolution.[7]

Clearly, these ventures represent very substantial investments. They are driven as much by competition between the two companies as by any demand from computer companies for improved chips. An additional dimension of competition flows from the jousting between these U.S. companies and foreign companies. So long as chip performance continues to improve, the U.S. companies can hold their lead. As soon as performance stops increasing, however, the foreign companies can catch up. Ultimately, computer CPU chips will become a commodity, as many kinds of memory chips already are, with the companies offering the lowest prices dominating the market.

The high cost of Intel's and AMD's factories shows that corporations may be willing to make big investments, but only in a well-established, competitive industry in which markets and technical characteristics evolve gradually. This would not be the case for nanoelectronics based on carbon nanotubes, however. A wholly new form of nanoelectronics would need to go through a long and costly period of development before it could compete in terms of cost, reliability, and performance with conventional computer chips. No one knows how long this phase would last or how much money would need to be invested, and this uncertainty makes the investment especially risky.

Mass production would probably require the creation of a suite of support industries, in addition to the factories that actually make the components equivalent to today's chips. One reason is that very different materials would be involved in this manufacturing process. Another reason is the high likelihood that very different circuit architectures and programming principles could be required. It is widely believed that components composed of carbon nanotubes would have a high failure rate, because they would be constructed at the absolute physical limits of manufacture and testing. As a consequence, the circuit would need to be adaptive, automatically working around dead transistors. At the same time, to achieve an acceptable degree of transistor reliability even with adaptive circuitry would require very high purity of raw materials and precise control over the manufacturing process; achieving this goal would not only be costly in terms of the manufacturing process itself but would also require large investments in the specialized support industries that provide the materials and the equipment for manufacturing and testing.

It is not clear what might justify the huge investment required to create an entirely new molecular computing industry. In the year 2005, computer scientists joked, "Moore's law ended two years ago." Indeed, the performance of the central processing chips in home and office PCs practically stopped increasing about that time, although other measures of performance continued to increase. Most home and office users do not need improved performance, with the possible exception of better Internet connectivity, to run current applications. Players of online games will need to choose between PCs with special graphics boards or the new generation of dedicated videogame systems, and they may ultimately find that neither is really as good as they want. Perhaps scientific, security, or other government-sponsored applications might demand a revolutionary shift in the design of computers. Another possible route to revolution could be the discovery of an entirely different kind of computer technology, whether nanoscale or not, arising from converging technologies.

Carlo Montemagno (Figure 3–1) believes that a new approach to computer design could be based on analogies with biological concepts.[8] One limitation of integrated circuits is the fact that the circuitry is essentially flat, with the components being laid out on a two-dimensional surface. Electrical engi-

Figure 3–1 Carlo Montemagno. Currently Dean of the College of Engineering of the University of Cincinnati, Dr. Montemagno is not only a leading researcher at the nano-bio-info intersection, but also played key roles in organizing NBIC conferences and promoting convergence throughout the science and engineering communities.

neers are aware of this limitation and occasionally design multilevel chips, but as Moore pointed out in his 1965 article, a two-dimensional chip has the advantage that it radiates the waste heat it produces. By contrast, brains, whether those of humans or animals, are three-dimensional. Montemagno imagines that it might become possible to assemble a three-dimensional computer, perhaps composed of bubble-like nanoscale vesicles with specially designed membranes that allow them to interact like transistors. A wholly new engineering approach may be needed to bring Montemagno's vision to reality, such that the three-dimensional hierarchical structure of vesicles would be allowed to emerge rather than being designed in from the start. In principle, there might be no limit to the size of such a unit, as it might conceivably include hundreds of billions of vesicles. Such a biomimetic computer might have less efficient circuitry and possibly slower individual components, but could greatly outperform today's computers through massive parallel computing.

Moore's law is not a law of nature, but rather is chiefly a historical description of the rate of improvement of integrated circuits over a remarkably long span of time. I have also seen it used as a normative rule—that is, as a statement that we *should* invest heavily enough to keep the technology improving at this rate. Sometimes software engineers and other computer scientists privately admit a different formulation: Unless the performance of integrated circuits continues to improve at something like the historical rate, improvement of software and information systems will soon cease. This point has a certain "political" quality within science and engineering, because computer scientists don't like to think they are dependent upon electrical engineers. But they are. An immediate corollary is an economic question: Who demands improved performance and is willing to pay for it?

SENSORS

If it is true that no current application demands a revolutionary shift toward some new form of nanocomputing, there remains a second route by which it could be developed. Nanotechnology may be essential for many valuable kinds of sensors—for example, chips that can quickly detect nerve gas in the air before it harms soldiers. As development of an increasing variety of sensor devices expands technical capabilities at the nanoscale, we might begin to see hybrid devices in sensornets. For example, a nerve-gas detecting chip could do all the necessary sensory and computational work, and then send a message across the soldiers' information system combined with input from dozens of similar chips. Soldiers would then know not only that nerve gas has been released, but also exactly where and in which direction they should evacuate. Over time, the advanced nanoelectronics developed for specialized, high-value sensors might be adopted for more uses, until ultimately they replaced silicon-based central processing chips.

In 2004, nanotechnology pioneer Paul Alivisatos argued that the coming decade would feature the development of nanocrystal-based methods for detecting single molecules in biological systems.[9] Long before nanosensors are applied to practical tasks in society, they will assist researchers in biochemistry. For example, Wayne Wang and associates at Harvard University have used silicon nanowire transistors to detect the biochemical processes implicated in a form of leukemia.[10]

Nanosensor research is highly convergent. For example, the Nano Sensors Group at the University of Illinois at Urbana–Champaign describes its work as follows: "The group is highly interdisciplinary, with expertise in the areas of microfabrication, nanotechnology, computer simulation, instrumentation, molecular biology, and cell biology. In particular, we are working on biosensors based upon photonic crystal concepts that can either be built from low-cost flexible plastic materials or integrated with semiconductor-based active devices, such as light sources and photodetectors, for high-performance integrated detection systems."[11]

A major recent development in computer science is the integration of large numbers of small sensors into sensornets, or sensor-based networks that may be distributed either across the land, across a production facility or other architectural structure, or across the human body. David Culler of the University of California at Berkeley, writing with Deborah Estrin and Mani Srivastava of the University of California at Los Angeles, says that the uses of sensornets can be classified into three categories—monitoring spaces, monitoring things, or monitoring the interactions between things and spaces:

> The first category includes environmental and habitat monitoring, precision agriculture, indoor climate control, surveillance, treaty verification, and intelligent alarms. The second includes structural monitoring, ecophysiology, condition-based equipment maintenance, medical diagnostics, and urban terrain mapping. The most dramatic applications involve monitoring complex interactions, including wildlife habitats, disaster management, emergency response, ubiquitous computing environments, asset tracking, health care, and manufacturing process control.[12]

Each unit in a sensornet consists of the sensor device itself, a transmitter–receiver with which to exchange data with the network, and a power supply of some kind. Nanotechnology enters the design in each of these areas.

Imagine the purpose of a sensornet is to protect the perimeter of a military facility by detecting chemical attacks. As we have seen, nanoscale sensors could identify molecules of the chemical agents. Popular terminology, however, sometimes oversimplifies the true nature of these substances. For example, we hear about the use of chlorine and phosgene gases in World War I, and the development of the nerve gases tabun and sarin by the Germans in World War II. Chlorine and phosgene are both genuine gases; by contrast, tabun and sarin are liquids that can act like gases when distributed in the form of a mist of liquid droplets or as an aerosol in combination with stabilizing chemicals. Chlorine gas consists simply of pairs of chlorine atoms, whereas each phosgene molecule contains two chlorine atoms, one carbon atom, and one oxygen atom. Tabun and sarin are more complex molecules, with their sizes being at the lower end of the nanoscale but much bigger than a two-atom chlorine molecule (i.e., Cl_2). The tabun molecule consists of 21 atoms of carbon, hydrogen, nitrogen, oxygen, and phosphorus, whereas the sarin molecule contains 18 atoms of carbon, hydrogen, fluorine, oxygen, and phosphorus. Chlorine has a powerful odor and immediately irritates human flesh. By comparison, humans find it much more difficult to detect phosgene, tabun, and sarin.

An advanced device for detecting chemical warfare agents would presumably carry out some computations on the data before transmitting them—for example, counting the abundances of the molecules and encoding the data. The smaller the onboard computer and the transmitter–receiver, the better, not only to make the entire sensor small enough to escape notice by an enemy soldier who happened to look in its direction, but also to reduce its power needs and thus permit a smaller or longer-lasting power supply. The range of a tiny transmitter will be short, a fact that has inspired many computer scientists to design networks of sensors that communicate with one another, rather than directly with a base station. In this way, the sensors combine their collective information to get a general picture of what is happening in the environment before the information is sent home either by the sensor that happens to be nearest or by specialized transmission devices with greater power. Nanotechnology may also be useful in building energetic power supplies, which could comprise tiny systems combining a solar cell with storage battery, or drawing power from intermittent microwave beams sent to them from the base station.

QUANTUM COMPUTING

One of the most challenging areas of computer science research related to nanotechnology is the quest to develop quantum computing. One way to think about quantum computing is in terms of the mathematical problem most often discussed as a possible early application of this technology, the factoring of large numbers. One of the most remarkable mathematical facts is that every positive integer can be expressed as the product of two prime numbers. For example, 6 is the product of 2 and 3. This is true for prime numbers as well, because each is the product of itself and 1—for example, 5 times 1 equals 5. By definition, a prime number has no integer factors except itself and 1. Over the years, mathematicians have exerted great effort to seek principles that describe primes as part of the larger activity called number theory, which has implications for other areas of mathematics.

Lifetimes have been spent seeking rules to construct larger prime numbers or to factor large numbers efficiently to see if they might be primes. Anyone can understand the straightforward method. Given a number—say, 41—you try to divide every prime number less than the square root of 41 into 41 and check whether the result is an integer. The square root of 41 is approximately 6.4. Setting 1 aside, the primes less than 6.4 are 2, 3, and 5. If you don't happen to have a list of all prime numbers less than the square root of 41, you may need to divide all the integers in that range into 41, unless a mathematical trick allows you to skip some of them. For example, you can skip all even numbers greater than 2, and you can spot them by the fact that their right-hand digit is also even. In the case of 41, that simplification trims the list of divisors to just 2, 3, and 5.

As you deal with increasingly larger numbers, you face an increasing amount of work to find the factors. For example, if you are checking whether 199 is a prime, the primes smaller than the square root are 2, 3, 5, 7, 11, and 13. If you don't have that list, you must try dividing 9 into 199 as well. Not counting 1, the one-thousandth prime is 7,919. To determine that 7,919 is a prime number, you would need to divide it by 44 integers (23 primes and 21 non-primes) given that you know the even-number trick but lack a list of primes.

Very large integers become practically impossible to factor, unless a trick like spotting that the integer is even happens to apply. Today, we use computers to handle this job, and the first 15 million primes can be downloaded from a website at the University of Texas at Martin.[13] At the moment, the largest known prime is 9,152,052 digits long.[14] This book has room to print only about 500,000 digits, so you would need 18 books this size to print all the digits of this number. However, this number is a *Mersenne prime*, which means that it is generated by taking 2 to a large prime exponent and then subtracting 1. In this case, the number is 1 less than 2 to the 30,402,457th power (i.e., $2^{30,402,457} - 1$). However, computers can use very efficient algorithms for testing whether a given Mersenne number is a prime, and many much smaller numbers are considerably tougher to factor.

Aside from their value in testing computer hardware and their curiosity value, what good are prime numbers? They have become very important in the specialized area of public key cryptography, a method of coding and decoding secret messages based on the fact that it can be much easier to multiply two prime numbers together than to factor the result.[15] Public key cryptography allows people to encode messages without having the ability to decode them, which gets around the problem that codes might be cracked by stealing a codebook from one of the people using it. Prime numbers are commonly used to code messages, including very brief messages that serve as electronic signatures verifying the identity of the sender of a message that itself is not encoded. Public key cryptography can be useful in online commerce as well as for private communications between corporations, governments, or terrorist groups. Thus governments could often find it useful to crack codes by factoring large numbers into the two primes that would allow them to read messages.

This is where quantum computing comes in, at least in theory. The level below the bytes of a digital computer consists of the binary bits. In binary arithmetic, the number 7 is represented as 111, which means $(1 \times 4) + (1 \times 2) + (1 \times 1) = 7$, because the 1s and 0s of binary represent powers of 2 rather than powers of 10 as in the decimal system. In a conventional computer, each of the bits at any given moment is either a 1 or a 0. In a hypothetical quantum computer, the numbers are stored in qubits (quantum bits), which are like bits, except that a quantum phenomenon called superposition allows a qubit to represent both 1 and 0 simultaneously. In regular computers, three bits can store only one of the numbers from 0 through 7 at any given time: 000, 001, 010, 011, 100, 101, 110, or 111. In principle, three qubits can store all eight numbers at the same time. When a regular computer uses brute force to factor a big number, it tries divisors one at a time. Hypothetically, a quantum computer could perform all the arithmetic in one fell swoop.

Suppose you had a quantum computer just big enough to store a number and its two prime factors. Conceivably, it could reside in a special chip in one corner of an ordinary laptop, where it would be used just for special tasks. Suppose you feed the quantum computer a large number, such as 275,604,541. Instantly, it divides the 16,601 integers less than the square root into this number and tells you its factors are 275,604,541 and 1, which means it is a prime. (Indeed, it is the 15,000,000th prime.) If you feed in 10,898,379,079,671,203 instead, the quantum computer instantly tells you the prime factors are 104,395,301 and 104,395,303.

Whether it will be possible to build functioning quantum computers and whether they might be able to accomplish other jobs remain unclear at the present time. However, scientists and funding agencies have a good deal of optimism about the potential of quantum computing. The 2004 *Quantum Information Science and Technology Roadmap* published by Los Alamos National Laboratory identified seven technically distinct approaches to this issue and acknowledges the possibility that others might be imagined.[16]

Quantum computing may or may not be a form of nanotechnology, depending on which technical approach is used. As commonly conceptualized, nanotechnology involves solid substances that maintain a distinctive structure at the nanoscale. Liquids, gases, and plasmas do not maintain a nanoscale structure in the same way that solids do. Research in nuclear magnetic resonance (NMR) quantum computation often works with liquids, for example. Thus the imaginary qubit component in the laptop mentioned previously might properly be called a *bottle* rather than a *chip*, if such a component becomes possible. It may be that quantum computing requires special conditions, such as a super-cold temperature, that can be sustained only in a large laboratory. We can imagine that at least one of the approaches to quantum computing will succeed under normal conditions of temperature, pressure, and other variables. Given that ordinary microelectronics is already working at the nanoscale, one would think there would be no reason to build quantum components any larger.

A very different model of quantum computing might emerge with other methods such as DNA computing (which lets a soup of molecules carrying information as a genetic code perform the calculations in parallel) or Montemagno's bubble computer. In this scenario, scientists identify a very difficult problem that needs to be solved through massive computations. A specialized computer is built, with its characteristics optimized for that one problem. It is run once, thereby solving the problem. Because it is incapable of solving any other problem, the computer is then junked. Indeed, the process of solving the problem incidentally ruins the quantum material or DNA or bubble-like nanoscale vesicles, so solving a problem inescapably means destroying the computer. This image of advanced computing suggests we may continue using conventional computer chips for all general-purpose computing. For all ordinary nonscientific applications, then, Moore's law will have reached the end of its life span, even as scientists keep improving the specialized computers they use in their own work.

THE INFORMATION TECHNOLOGY RESEARCH INITIATIVE

Central processing unit chips are merely one part of information technology, and massive efforts have been invested in developing the full range of components, methods, and applications. Perhaps the best single example of a coherent program to advance information technology that has implications for other fields of science and technology is the multi-agency U.S. government initiative currently called Networking and Information Technology Research and Development, spearheaded by the 1999–2003 Information Technology Research (ITR) Initiative at the National Science Foundation (NSF).[17] Two separate developments came together to make ITR possible: the Knowledge and Distributed Intelligence (KDI) program and the President's Information Technology Advisory Committee (PITAC).

Knowledge and Distributed Intelligence was a two-year precursor to the Information Technology Research Initiative. Carried out by NSF in 1998–1999, KDI linked computer science with biology, chemistry, physics, mathematics, geosciences, engineering, and social sciences. Unlike the later ITR, it gave special emphasis to cognitive science under the rubric of "learning and intelligent systems." As explained in the announcement for the first grant competition, the aims of KDI were "to achieve, across the scientific and engineering communities, the next generation of human capability to generate, gather, model, and represent more complex and cross-disciplinary scientific data from new sources and at enormously varying scales; to transform this information into knowledge by combining, classifying, and analyzing it in new ways; to deepen our understanding of the cognitive, ethical, educational, legal, and social implications of new types of interactivity; and to collaborate in sharing this knowledge and working together interactively."[18] KDI was based on four principles that paved the way for converging technologies:[19]

1. To create a ubiquitous information environment that can provide needed knowledge to anyone, anywhere, anytime.

2. To understand the nature of human, animal, and machine intelligence.

3. To drive economic growth on the basis of new scientific knowledge and technological capabilities.

4. To build knowledge environments for science, engineering, and education based on multidisciplinary collaborations.

After the two KDI competitions were complete and enough time had passed for the research projects to achieve their results, a retrospective study was done. A total of 71 KDI projects were funded, averaging about $1.5 million each. The report completed as part of this study explained the basic philosophy of KDI: "Multidisciplinary projects can foster invention, the development of new areas of inquiry, and the development of careers in the frontiers of science and engineering."[20] Collaboration across fields and cooperation between different institutions demanded coordination through active supervision, direct communication in regular meetings and special events, and a significant amount of travel even in the age of Internet. Perhaps the most important lesson of KDI was that efforts intended to create multidisciplinary teams to address new challenges were usually successful, as measured by peer-reviewed publications and new technologies.

On February 24, 1999, the President's Information Technology Advisory Committee submitted a report to President Bill Clinton, titled *Information Technology Research: Investing in Our Future*, arguing that current levels of federal support were inadequate. This report stressed the transformative nature of information technology, but described the benefits largely in terms of their ability to improve the conditions under which citizens live and work, rather than suggesting information technology would fundamentally change the structure of society or the nature of humans:

Information technology will be one of the key factors driving progress in the 21st century—it will transform the way we live, learn, work, and play. Advances in computing and communications technology will create a new infrastructure for business, scientific research, and social interaction. This expanding infrastructure will provide us with new tools for communicating throughout the world and for acquiring knowledge and insight from information. Information technology will help us understand how we affect the natural environment and how best to protect it. It will provide a vehicle for economic growth. Information technology will make the workplace more rewarding, improve the quality of health care, and make government more responsive and accessible to the needs of our citizens.[21]

Much of the PITAC report focused on continued progress in the traditional computer science areas of software research and high-end computing (or supercomputing). The supercomputer industry had experienced a serious shakeout in the mid-1990s, and the PITAC report was a welcome testimony to the continuing value of machines that were very large and fast. Among its technical research priorities were two somewhat less traditional topics: scalable information infrastructure and research on the socioeconomic implications of information technology. Missing from the PITAC report, however, were any references to cognitive science or convergence between computer science and the sciences of human intelligence.

In 1999, NSF issued the first of five annual competition announcements for the ITR initiative, which had a wider scope than the PITAC recommendations but did not include the cognitive science emphasis area that had been important in KDI.[22] In this first year of an intended five-year lifetime, ITR emphasized work in the following areas:

- Software engineering
- Scalable information infrastructure
- Human–computer interface
- Information management
- Advanced computational science
- Revolutionary computing
- Social and economic implications of information technology
- Information technology education and workforce

In its second year, ITR expanded to include applications of information technology across the sciences and engineering.[23] The third competition expanded the scope even further, "to enable research and education in multi-disciplinary areas, focusing on emerging opportunities at the interfaces between information technology and other disciplines."[24] Thus, in the very middle of its duration, ITR emphasized technological convergence. The fourth and fifth years of the competition continued to balance fundamental computer science research against applications across the sciences, and in the wake of the terrorist attacks of September 11, 2001, they added a security dimension.[25] Over the five years, NSF announced it had invested $689 million in ITR.[26] After the fifth and final year of the special ITR competitions, I searched through all of the awards using the publicly available NSF abstracts database, looking for projects that particularly promoted convergence.[27] My report, "Information Technology for Convergence," was published in the third *Converging Technologies* volume.[28]

Several of the ITR projects were in the new domain that might be called *computational nanoscience,* in which conventional large computers advance the goals of nanotechnology. Several of these projects used computer simulation to model physical processes at the nanoscale. For example, a project headed by Rodney Bartlett at the University of Florida modeled the effect of water on nanostructures; Jonathan Freund at the University of Illinois at Urbana–Champaign modeled the effect of heat; and a collaboration between Lizhi Sun at the University of Iowa and Nasr Ghoniem at UCLA modeled deformation and dislocation. Other projects sought to develop methods for simulating how phenomena at the nanoscale relate to phenomena at larger scales, including work by John Bassani at the University of Pennsylvania and a collaboration among Yannis Kevrekidis at Princeton, Dimitrios Maroudas at the University of California at Santa Barbara, Robert Armstrong at MIT, and Mark Swihart at the State University of New York at Buffalo. Simulations like these typically model the positions of individual atoms in a nanoscale structure, and assess how they dynamically influence one another as the structure grows through the addition of more atoms or changes in response to external factors.

Several other ITR projects focused on the field of *nanocomputing,* developing the technologies to extend Moore's law through progress toward molecular electronics. Alvin Lebeck at Duke University received funding so that he could explore the possibility of using DNA to assemble nanoscale building blocks such as carbon nanotubes into nanoscale electronic circuits. Junichiro Kono at Rice University, Christopher Stanton at the University of Florida, and Lu Sham at the University of California at San Diego explored optical methods for controlling the properties of materials for nanoelectronics and spintronics. Whereas electronics is based on the charge of an electron, spintronics is based on the electron's spin or magnetic characteristic.[29] ITR support for the Institute for the Theory of Advanced Materials in Information Technology, spanning the Universities of Texas and Minnesota, enabled research on nanowires, spintronics, and quantum dots—that is, semiconductor nanocrystals with unusual optical and electric properties.[30]

Another ITR nanocomputing grant went to MIT's Center for Bits and Atoms, a visionary research group that describes itself as "an ambitious interdisciplinary initiative that is looking beyond the end of the Digital Revolution to ask how a functional description of a system can be embodied in, and abstracted from, a physical form."[31] This organization seeks to conceptualize objects in terms of their information content, which can be divorced from the physical material the object is composed of. Members plan to use computer descriptions of objects to reproduce them from many different materials at a range of sizes, perhaps all the way down to the nanoscale. However, the most famous product of the Center for Bits and Atoms is its international network of "fab labs," desktop computer-controlled fabrication systems that operate at a scale much larger than the nanoscale and allow students to make a variety of objects from data specifications.

A number of ITR projects involved quantum nanotechnology, in which researchers try to exploit quantum effects in nanoscale structures, such as quantum dots, which may become fundamental components of future computing systems. Quantum effects begin to become significant over distances of a few nanometers, and the quanta themselves are subatomic and thus smaller than a nanometer. However, quantum effects can be a nuisance for nanotechnology—for example, introducing noise into nanoelectronic circuits—and must be fully understood if these effects are to be managed successfully. John Preskill of the California Institute of Technology received an ITR grant funding his work related to the suppression of unwanted quantum effects in nanoscale electronic devices, and Thomas Beck of the University of Cincinnati got one to explore quantum effects on the movement of electrons through molecular wires.

Other ITR convergence research focused on sensors to detect and identify chemical hazards, nanoscale particles, and microorganisms. Several projects focused on the area that might be called nanoscale bioinformatics—information systems devoted to proteins, DNA sequences, and other small structures in living organisms. Naturally, several ITR projects advanced biotechnology, and some of the new cognitive technologies offered computational neuroscience tools for studying the brain and developed educational technologies. Biocomputing included calculating by means of DNA or designing computers by analogy with living systems (which will be explained further in Chapter 4). In addition, a few projects focused on human–technology interaction, including the societal impacts of convergent information technology.

GRAND INFORMATION TECHNOLOGY CHALLENGES

In the wake of ITR during 2003, I was privileged to serve on the Grand Challenges Task Force of the U.S. government's Interagency Working Group on Information Technology Research and Development. Eight years earlier, a similar group had published a set of grand challenges for high-performance computing and communications, and it was time for a fresh vision of the future. Participants included information technology experts from U.S. government agencies: the Agency for Healthcare Research and Quality, the Defense Advanced Research Projects Agency, the Department of Energy's Office of Science, the Environmental Protection Agency, the Federal Aviation Administration, the National Institutes of Health, the National Institute of Standards and Technology, the National Oceanic and Atmospheric Administration, the National Security Agency, the National Science Foundation, the Office of the Deputy Undersecretary of Defense for Science and Technology, and the National Science and Technology Council; their work was managed by the National Coordination Office for Information Technology Research and Development.[32]

The Grand Challenges Task Force identified sixteen grand challenges that together could provide leadership in science and technology that would benefit national security, health and the environment, economic prosperity, a well-educated populace, and a vibrant civil society. I will not try to summarize the report in this section, but instead will show how each of the sixteen grand challenges involves technological convergence and could enhance human well-being, based on my own understanding of the challenge and the relevant information technology.

1. Nanoscale Science and Technology: Explore and Exploit the Behavior of Ensembles of Atoms and Molecules. To design and control the manufacture of nanoscale devices and complex structures, we will need ultra-fast computers that are able to calculate the exact behavior of atoms and molecules in any desired combination and configuration. At present, laboratory research on many nanoscale materials is rather haphazard, trying various procedures and comparing the results of many attempts. In the future, however, it will be possible to calculate everything with high precision, on the basis of our knowledge of the laws by which atoms are assembled into molecules, composites, and nanoscale machines. The task force judged that success in this effort could achieve a Second Industrial Revolution, transforming manufacturing as radically as the original Industrial Revolution did two centuries ago.

2. Clean Energy Production Through Improved Combustion. Humans have controlled fire for 300,000 years, yet we still have much to learn about how the combustion process operates. Very complex but highly accurate computer models will be needed to model combustion—for example, the reactions of gasoline in internal combustion engines. This means modeling everything from the nanoscale movement of the molecules of gasoline vapor, to the constantly changing temperature of every small region of gas inside the engine, to the periodic variations in the size and shape of the combustion chamber itself. The goal is to be able to control the chemical reaction and physical forces so precisely that the maximum useful energy could be extracted with minimum fuel consumption and minimum pollution generation.

3. Knowledge Environments for Science and Engineering. The task force noted that collaboratories already exist to support research on the upper atmosphere, earthquakes, biomedical problems, and astronomy. A *collaboratory* is a system that allows scientists, engineers, and students at remote locations to communicate with one another, to access shared databases and analysis software, and sometimes to operate instrumentation for gathering new knowledge and machinery for creating design prototypes of new technology. The report noted that the needs of scientists are changing, given that "data sets are more complex and teams are more interdisciplinary," and predicted that a major benefit of advanced knowledge environments would be "new discoveries across disciplines (for example, discoveries in one field can apply to other fields)."[33]

4. High-Confidence Infrastructure Control Systems. In current information systems lingo, "high confidence" is the buzzword for reliability, referring to systems that seldom break down and that have effective backups that smoothly kick in on those rare occasions when something does go wrong with the main system. In this context, "infrastructure" refers to utilities such as the electric power grid, an ample supply of clean water, and the transportation system consisting of boats, planes, trains, trucks, and automobiles. The essential infrastructure might potentially be attacked by terrorists, be disrupted by natural disasters, wear out, or suffer outages when complex failures cascade from a local breakdown to affect the entire system. One way to deal with this last problem might be to apply knowledge gained through operation of Internet, which was originally designed to be a distributed system that could not be halted by any local problem.

5. A Safer, More Secure, More Efficient, Higher-Capacity, Multimodal Transportation System. The transportation system deserves special mention, because it is a system of systems, only some parts of which (e.g., air traffic control) are unified infrastructures that can be managed by a single information system. Convergence enters the picture at many points, including the connection of different forms of transport in the most effective way, the development of safer vehicle control based on cognitive science (of drivers and pilots) and information technology (e.g., computer vision collision avoidance), and the need for new materials, such as nanostructured composites for stronger but lighter vehicles. Benefits would include saving time for travelers, improved safety, and economic efficiency.

6. Real-Time Detection, Assessment, and Response to Natural or Human-Made Threats. Quick, correct information is fundamental to dealing with modern threats. Suppose a chemical or biological agent were released inside a large office building. Ideally, the ventilation system of the building would automatically identify the threat, notify first responders, and adjust the flow of air to contain the threat as much as possible. To protect lives and property, instantaneous warnings of earthquakes and high-accuracy, real-time predictions of the local effects of hurricanes could be extremely valuable. Distributed sensor systems, which monitored changing conditions over many time scales, would feed information into constantly running computer simulations, which in turn would direct traffic flow, dispatch rescue teams, and even send robots for removal of hazardous materials.

7. Improved Patient Safety and Health Quality. Medical informatics operates at the boundary of biotechnology and information technology. It increasingly makes use of findings from cognitive science, especially in the area of human–computer interaction, because information is of no use unless physicians and other health caregivers can apply it effectively. Today, many medical decisions are based on poor understanding both of the condition of the patient and of the objective value of the available treatments. The current popularity of the term "evidence-based medicine" underscores the need for scientific information about treatments, which would include these therapies' effects on different kinds of patients, including any genetic susceptibilities to side effects. All too often, information is missing about a patient's medical history. Collecting, preserving, and intelligently using more data about patients would not only provide them with better care, but also build a shared knowledgebase about the values of treatments and the causes of errors.

8. Informed Strategic Planning for Long-Term Regional Climate Change. Awareness is growing about the issue of global warming, but people may not be aware that climate change will affect different regions in potentially different ways. Agricultural regions may experience increased frequency of droughts, shorelands may disappear beneath the sea, and while most regions are warming a few may get colder because of alterations in ocean currents. Much of the research in this area will involve supercomputer simulations, starting with huge data sets describing weather and climate for large numbers of locations, modeling future trends, and comparing results with actual data to improve the computer models and the theories underlying them. At the same time that climate science is improving in this iterative manner, the best continually updated assessments will be provided to policy makers so that societies can adapt to changes, mitigate their effects, and generally improve the quality of their decision making.

9. Predicting Pathways and Health Effects of Pollutants. Understanding the health effects of pollutants requires convergence of environmental, biological, and medical sciences. For example, nanoscale particles can have distinctly different chemical and biological characteristics from bulk forms of the same materials, largely because of their greatly increased surface areas (and thus greater rapidity of chemical action) and their ability to invade small spaces (such as the tissues of the human body). Specific empirical studies will be needed to determine how nanoparticles are transported by air or water in the natural environment, how they cross barriers such as human skin, and how they affect living cells or the DNA that carries the human genetic code. Advanced forms of computation will be required to carry out each specific study, but then an integrated information system will be needed to model how the various steps link together to give various particulate substances different degrees of actual (as opposed to hypothetical) hazardousness.

10. Managing Knowledge-Intensive Organizations in Dynamic Environments. In a constantly changing world, organizations must be agile, responding quickly and appropriately to both threats and opportunities. At the same time, decisions seem to be growing more complex, and ever greater numbers of factors need to be considered before choices are made. For example, a manufacturing corporation needs to be able to change the specifications of its products on short notice in response to changing markets, even as its supply chains are in constant flux in a global business environment. Among the computational methods available for modeling complex social organizations are multiagent systems, in which large numbers of artificial intelligence units interact according to evolving models. Computers alone are not enough to permit constant reconfiguration of processes and redeployment of resources, because everything from materials to the company's design philosophy to the very organization of the workforce will need to become more flexible. The organization itself must become the constant focus of research, as lessons learned are fed back into the system rapidly and used to manage the organization for success in a dynamic world.

11. Anticipating the Consequences of Universal Participation in a Digital Society. In addition to the many benefits that the creators of information technology have intended, inevitably some unintended problems—such as identity theft or inequality of access to information—must be addressed. While computer hardware has constantly improved at an exponential rate, the software engineering to serve human needs has failed to advance equally rapidly. Cognitive research will help by focusing on computer designers as well as users, as will social research focusing on the effects of new technologies. Special research efforts will be needed to understand how best to serve disabled people and other individuals who are currently left out of the information society for cultural or economic reasons, and to recognize when new public policies concerning information technology are needed.

12. Generating Insights from Information at Your Fingertips. Progress on many of the other grand challenges will require new and better ways to extract meaningful information from vast floods of apparently meaningless data. Cognitive science will contribute to this effort by identifying human mental biases and suggesting ways to overcome them. Networked computers can collect, manage, and distribute data. We will need improved software tools to locate needed information, translate information across formats and modalities, find connections between seemingly separate facts, classify data by means of appropriate taxonomies, and automate the creation of metadata descriptions of freshly collected data. Success will accelerate scientific progress, speed up wise decision making, and promote cooperation across disciplines and communities.

13. Collaborative Intelligence: Integrating Humans with Intelligent Technologies. This challenge directly concerns convergence of information and cognitive technologies, with a special focus on building collaborative teams composed of people, software agents, robots, and sensors. Software agents are relatively autonomous programs that do jobs for the benefit of people and that operate inside computers, distributed information systems, or robots. When their work involves rapidly changing circumstances, the agents and robots need considerable autonomy, even as they take direction from humans. Applications include rapid response to disasters, increased productivity in the services sector of the economy, and agile manufacturing that offers the economic advantages of mass production and the quality advantages of customization.

14. Rapidly Acquiring Proficiency in Natural Languages. This challenge would greatly advance the information technology area called "natural language processing" through research on how humans learn language, whose results might then be applied to computational models to improve the language education of humans. It explicitly involves modeling human cognitive processes and their relationship to the structure of language. A wide range of experiments would be done "with different learning models (such as reinforcement, evolutionary, clustering, and supervised)," and ultimately partial models would be merged into a cohesive total model of human and machine language.[34] The results would greatly improve automatic systems for searching written text and recorded speech for desired information as well as lead to the design of better methods for teaching language skills to children, immigrants, and anyone who needs to learn an additional language.

15. SimUniverse: Learning by Exploring. A suite of educational computer simulations, created with advanced computing technologies but playable on widely available machines, would give students at many levels the opportunity to learn by experimenting with realistic models of natural phenomena. Biological simulations could demonstrate how the human digestive system extracts nutrients, how the circulatory system distributes them through the body, and how the immune system responds to disease. Simulations of the Earth's atmosphere could teach students about seasonal weather patterns, global climate change, or the effects of air pollution on the environment. Astronomical simulations could display the changing dynamics of the solar system from its origins over any desired scale of time or distance. The modular approach would be coupled with excellent designer and user interfaces, so that teachers in any field could create their own simulations.

16. Virtual Lifetime Tutor for All. The tutor envisioned in this grand challenge would be a personalized information system with artificial intelligence, which would adapt to an individual user's changing educational needs over the entire span of life, from birth to old age. As the user's needs change, the tutor could emphasize different subjects—for example, just-in-time technical training in nanotechnology when the user's employer began to incorporate new nano methods in the individual's work. A person who wanted to learn a language such as Spanish could start with simple lessons in his or her spare time; the tutor would then accompany its owner on a trip to a Spanish-speaking country, acting as interpreter when needed but always seeking to help the human student learn. While the tutor could use curricula in any of the NBIC areas, its design, and its ability to understand its owner, would require advances in cognitive science.

Several of these sixteen challenges connect information technology to other NBIC fields to achieve particular applications, and some will serve as the basis of entirely new manufacturing and service industries. Crucially, the majority seek to revolutionize the way research and education are done across all fields of science and engineering, through new forms of informatics tools and knowledge infrastructure, in what can be called *cyberinfrastructure*.

CONCLUSION

Several of the sixteen grand challenges described in the preceding section require new computing and communication facilities—that is, development of cyberinfrastructure. A special report on cyberinfrastructure posted on the NSF website calls it "a grand convergence" and notes the combined revolutionary impact of Internet, computer chips, and scientific databases: "The convergence of these and other trends have led researchers to envision a tightly integrated, planet-wide grid of computing, information, networking and sensor resources—a grid that we could one day tap as easily as we now use a power socket or a water faucet."[35] Cyberinfrastructure could be described as the marriage of supercomputers with digital libraries, plus much more.

Beginning with the construction of ENIAC at the end of World War II, many governments have invested in the largest computers feasible at the given time, with the goal of carrying out defense-related engineering calculations. ENIAC, which could be described as the first supercomputer ("super" in comparison with the calculating machines of its day), did work for artillery ballistics and nuclear weapons, for example. By the early 1980s, it was clear that supercomputers could accomplish much for some fields of civilian science, such as fusion energy, atmospheric sciences, and aerodynamics.[36] In 1985 and 1986, NSF established five supercomputer centers, at Cornell University, the University of Illinois at Urbana–Champaign, Carnegie Mellon University, the University of California at San Diego, and Princeton University.

When I joined NSF in mid-1992, one of the first special activities I undertook was organizing a workshop at the Illinois supercomputer center on what we called artificial social intelligence but today might be called multi-agent systems.[37] In 1995, I helped the social sciences division of NSF organize a workshop in connection with the supercomputer center at the University of California at San Diego, dealing with ways to enhance multidisciplinary scientific collaboration via Internet.[38] That same year, the Hayes Report advocated a fresh approach to organizing scientific supercomputing, and two years later NSF replaced the supercomputer centers with two Partnerships for Advanced Computational Infrastructure (PACI).[39] Coincidentally, these ventures were centered at the Illinois and California institutions with which I had worked previously. Each PACI consisted of a supercomputer center connected to approximately 50 other resourceful institutions in the form of a network. Thus it served as a model for distributed "grid" computing and for use of Internet to unite a scientific community.

As mentioned earlier, digital libraries will be a key component of the cyberinfrastructure. Digital libraries are large collections of knowledge, including images and multimedia objects, housed in information systems with tools for archiving, managing, and retrieving the contents for multiple users and multiple uses. The Digital Library Initiative (DLI) went through several phases before it was absorbed into other activities, as generally happens with the limited-duration activities called "initiatives." Beginning in 1993, DLI comprised a collaboration between the National Science Foundation, the Defense Advanced Projects Agency, and the National Aeronautics and Space Administration; in 1998, these parties were joined by National Library of Medicine, the Library of Congress, and the National Endowment for the Humanities.[40] Some of the best examples of the DLI's achievements involve historical rather than scientific data—notably, the website of the Library of Congress itself and the Perseus Project dedicated to ancient Greek and Roman culture at Tufts University.[41] Among the scientific examples are the international protein databases, the National Virtual Observatory in astronomy, and the DigiMorph library of images of natural history specimens such as vertebrate skeletons.[42] In 2001, PITAC recognized the vast importance of digital libraries, and the need for research to continue so that their scope and usability would continue to increase.[43]

The 2003 Atkins Report, expressing in detail the analysis of NSF's Advisory Panel on Cyberinfrastructure, was titled *Revolutionizing Science and Engineering through Cyberinfrastructure,* and it enthusiastically proclaimed that "a new age has dawned."[44] The revolutionary idea was not merely that information technology would enable rapid progress across all areas of science and engineering, but also that it would transform the very ways these fields were organized, bringing people together in collaboration across boundaries of academic discipline and geographic distance.

REFERENCES

1. ftp://download.intel.com/museum/Moores_Law/Printed_Materials/Moores_Law_Perspective.pdf

2. Gordon E. Moore, "Cramming More Components onto Integrated Circuits," *Electronics,* 38(8):115, 1965.

3. Gordon E. Moore, "Progress in Digital Integrated Electronics," *Technical Digest 1975,* International Electron Devices Meeting, IEEE, 1975, p. 11.

4. Intel fact sheet, "Moore's Law: Raising the Bar," ftp://download.intel.com/museum/Moores_Law/Printed_Materials/Moores_Law_Backgrounder.pdf

5. Intel press release, "Intel 45nm Technology Will Take Future Platforms to New Performance-per-Watt Levels," January 26, 2006, http://www.intel.com/technology/silicon/new_45nm_silicon.htm

6. Intel news release, "Intel Opens Third High-Volume 65nm Manufacturing Facility: Intel Extends Leadership in Advanced Chip Manufacturing with Europe's First 65nm Chip Factory in High-Volume Production," http://www.intel.com/pressroom/archive/releases/20060622corp.htm

7. AMD press release, "Governor Pataki and AMD's CEO Ruiz Announce Plans for Multi-billion Dollar 300 mm Semiconductor Manufacturing Plant," June 23, 2006, http://www.amd.com/us-en/Corporate/VirtualPressRoom/0,,51_104_543~110287,00.html

8. Carlo D. Montemagno, "Integrative Technology for the Twenty-First Century," in Mihail C. Roco and Carlo D. Montemagno (eds.), *The Coevolution of Human Potential and Converging Technologies* (New York: New York Academy of Sciences, 2004, pp. 38–49).

9. Paul Alivisatos, "The Use of Nanocrystals in Biological Detection," *Nature Biotechnology,* 22:47–52, 2004.

10. Wayne U. Wang, Chuo Chen, Keng-hui Lin, Ying Fang, and Charles M. Lieber, "Label-Free Detection of Small-Molecule–Protein Interactions by Using Nanowire Nanosensors," *Proceedings of the National Academy of Sciences,* 102(9):3208–3212, 2005.

11. http://nano.ece.uiuc.edu/research/index.html; retrieved July 15, 2006.

12. David Culler, Deborah Estrin, and Mani Srivastava, "Overview of Sensor Networks," *Computer,* 37(8):42, 2004.

13. http://primes.utm.edu/

14. http://www.mersenne.org/30402457.htm

15. Ronald L. Rivest, Adi Shamir, and Leonard M. Adleman, "Cryptographic Communications System and Method," patent 4,405,829, U.S. Patent and Trademark Office, September 20, 1983.

16. Quantum Information Science and Technology Experts Panel, *A Quantum Information Science and Technology Roadmap* (Los Alamos, NM: Los Alamos National Laboratory, 2004), http://qist.lanl.gov/qcomp_map.shtml

17. http://www.nitrd.gov/

18. National Science Foundation, *KDI: Knowledge and Distributed Intelligence in the Information Age,* proposal solicitation NSF 98-55, http://www.nsf.gov/pubs/1998/nsf9855/nsf9855.pdf; cf. *KDI: Knowledge and Distributed Intelligence in the Information Age,* revised proposal solicitation, NSF 99-29, http://www.nsf.gov/pubs/1999/nsf9929/nsf9929.pdf

19. John Adam and Ellen Weir, *KDI: Knowledge and Distributed Intelligence* (Arlington, VA: National Science Foundation, 1998), http://www.nsf.gov/od/lpa/news/publicat/nsf9860/start.htm

20. Jonathan Cummings and Sara Kiesler, *KDI Initiative: Multidisciplinary Scientific Collaborations* (Arlington, VA: National Science Foundation, 2003, p. 5).

21. Bill Joy and Ken Kennedy (eds.), *Information Technology Research: Investing in Our Future* (Arlington, VA: National Coordination Office for Computing, Information, and Communications, 1999, p. 1).

22. National Science Foundation, *Information Technology Research (ITR),* program solicitation NSF 99-167, http://www.nsf.gov/pubs/1999/nsf99167/nsf99167.htm

23. National Science Foundation, *Information Technology Research,* program solicitation NSF 00-126, http://www.nsf.gov/pubs/2000/nsf00126/nsf00126.htm

24. National Science Foundation, *Information Technology Research (ITR)*, program solicitation NSF 01-149, http://www.nsf.gov/pubs/2001/nsf01149/nsf01149.htm

25. National Science Foundation, *Information Technology Research (ITR) Fiscal Year 2003 Announcement*, program solicitation NSF-02-168, http://www.nsf.gov/pubs/2002/nsf02168/nsf02168.htm; *Information Technology Research for National Priorities (ITR) Fiscal Year 2004 Announcement*, program solicitation NSF 04-012, http://www.nsf.gov/pubs/2004/nsf04012/nsf04012.htm

26. http://www.nsf.gov/od/lpa/news/press/00/pr0061.htm; http://www.nsf.gov/od/lpa/news/press/01/pr0174.htm; http://www.nsf.gov/od/lpa/news/02/pr0278.htm; http://www.nsf.gov/od/lpa/news/03/pr03103.htm; http://www.nsf.gov/news/news_summ.jsp?cntn_id=100449

27. http://www.nsf.gov/awardsearch/

28. William Sims Bainbridge, "Information Technology for Convergence," in William Sims Bainbridge and Mihail C. Roco (eds.), *Managing Nano-Bio-Info-Cogno Innovations: Converging Technologies in Society* (Berlin: Springer, 2006, pp. 347–368).

29. Igor Zutic, Jaroslav Fabian, and S. Das Sarma, "Spintronics: Fundamentals and Applications," *Reviews of Modern Physics*, 76:323–410, 2004.

30. http://www.ices.utexas.edu/ccm/itamit/index.html

31. http://cba.mit.edu/about/index.html

32. Interagency Working Group on Information Technology Research and Development, *Grand Challenges: Science, Engineering, and Societal Advances Requiring Networking and Information Technology Research and Development* (Arlington, VA: National Coordination Office for Information Technology Research and Development, 2003).

33. Interagency Working Group on Information Technology Research and Development, *Grand Challenges: Science, Engineering, and Societal Advances Requiring Networking and Information Technology Research and Development* (Arlington, VA: National Coordination Office for Information Technology Research and Development, 2003, p. 12).

34. Interagency Working Group on Information Technology Research and Development, *Grand Challenges: Science, Engineering, and Societal Advances Requiring Networking and Information Technology Research and Development* (Arlington, VA: National Coordination Office for Information Technology Research and Development, 2003, p. 38).

35. David Hart, "Cyberinfrastructure: A Special Report," http://www.nsf.gov/news/special_reports/cyber/agrand.jsp

36. Peter D. Lax (ed.), *Report of the Panel on Large Scale Computing in Science and Engineering*, unpublished document, sponsored by the U.S. Department of Defense and the National Science Foundation in cooperation with the Department of Energy and the National Aeronautics and Space Administration, December 26, 1982.

37. William Sims Bainbridge, Edward E. Brent, Kathleen Carley, David R. Heise, Michael W. Macy, Barry Markovsky, and John Skvoretz, "Artificial Social Intelligence," *Annual Review of Sociology*, 20:407–436, 1994.

38. Alan Kornberg and Peter Arzberger, *Connecting and Collaborating: Issues for the Sciences*, http://www.sdsc.edu/CC/cc.html

39. Edward F. Hayes, *Report of the Task Force on the Future of the NSF Supercomputer Centers Program*, http://www.pnl.gov/scales/docs/nsf_centers_1995.pdf

40. Stephen M. Griffin, "NSF/DARPA/NASA Digital Libraries Initiative: A Program Manager's Perspective," *D-Lib Magazine*, July/August 1998, http://www.dlib.org/dlib/july98/07griffin.html

41. http://www.loc.gov/index.html; http://www.perseus.tufts.edu/

42. http://www.pir.uniprot.org/; http://www.us-vo.org/; http://www.digimorph.org/

43. Raj Reddy and Irving Wladawsky-Berger (eds.), *Digital Libraries: Universal Access to Human Knowledge* (Arlington, VA: National Coordination Office for Information Technology Research and Development, 2001).

44. Daniel E. Atkins (ed.), *Revolutionizing Science and Engineering through Cyberinfrastructure: Report of the National Science Foundation Blue-Ribbon Advisory Panel on Cyberinfrastructure* (Arlington, VA: National Science Foundation, 2003, p. ES1).

Chapter 4

Biotechnology

Probably the most visible and active convergence occurring in the borderland between two major fields is that uniting biotechnology with nanotechnology: bio-nano or nano-bio convergence. Arguably, biotechnology is the oldest of the four NBIC fields, dating back to prehistoric days when plants and animals were first domesticated, when techniques such as fermenting wine and culturing cheese were originally invented, and when a variety of more or less effective medical techniques were first systematized. Perhaps the very first application of biotechnology was the cooking of food, which may have begun as early as a half million years ago. By contrast, the debate regarding nanotechnology and cognitive technology takes a vastly different tack—namely, which of them is younger, as both have arisen as recently as our own lifetimes.

The fusion of nanotechnology with biotechnology is a momentous connection combining innovations from the entire sweep of human history with our radical new ability to understand and manipulate matter on the nanoscale. The nanoscale is where complex chemicals and life itself arise.

NANOTECHNOLOGY FROM THE PERSPECTIVE OF BIOLOGY

John Marburger, director of the U.S. Office of Science and Technology Policy, has joked that much biotechnology should be called *wet nanotechnology* because it "deals with the life processes in humans and other organisms at the molecular and nanoscale levels."[1] By comparison, the major government report *Nanobiotechnology* says that the National Institutes of Health (NIH) defines nanotechnology more narrowly:

> While much of biology is grounded in nanoscale phenomena, NIH has not reclassified most of its basic research portfolio as nanotechnology. Studies are classified as nanotechnology projects if they (a) use nanotechnology tools and concepts to study biology

or develop medical interventions, (b) propose to engineer biological molecules toward functions very different from those they have in nature, or (c) manipulate biological systems using nanotechnology tools rather than synthetic chemical or biochemical approaches that have been used for years in the biology research community.[2]

The work behind this 2005 report began in an NNI workshop that was held October 9–11, 2003, and was jointly sponsored by NIH and NSF. Although these two agencies often cooperate, it is worth noting the boundaries that define their unique territories.

Founded in 1887, NIH consists of 27 institutes and centers, each conducting its own in-house research while also supporting research in universities, medical schools, hospitals, and other institutions. Its mission is to understand the causes of diseases and other health conditions, find ways of preventing illness, develop effective treatments and cures, and serve as a source of health-related information.

NSF was founded much more recently, in 1950, and its *Grant Proposal Guide* says that "Research with disease-related goals, including work on the etiology, diagnosis or treatment of physical or mental disease, abnormality, or malfunction in human beings or animals, is normally not supported."[3] However, NSF's Directorate for Biological Sciences supports fundamental research in human biology (plus plant and animal biology), and other directorates often play supportive roles in research that ultimately leads to medical advances. For example, the NSF computer science directorate sometimes supports fundamental research in medical informatics.

Nanobiotechnology identifies four major areas of research opportunities that are likely to be explored in the next several years. First, advanced imaging technologies (e.g., atomic force microscopes) will allow researchers to see what is happening on the molecular level and will improve our knowledge of how nanoscale biosystems function, potentially suggesting ways to optimize clinical therapies. Second, nanoprobes and nanoscale genomic or proteomic analysis can observe the processes of life within the living cell. Third, nanoengineering techniques will help us understand biological nanosystems by assembling them outside living organisms, thereby allowing us to test whether we really understand the molecular mechanisms and potentially to develop useful devices based on their principles. Fourth, nanomedicine is an emerging field that promises to deliver medically valuable materials, devices, and diagnostic methods.

NANO-BIO CONVERGENCE

One way to obtain a bird's-eye view of nano-bio convergence is to glance at nanotechnology-related grants made by the NSF Biological Sciences directorate. The online public database of NSF grants lists 44 explicitly connected to nanotechnology made through September 2006, for a total investment of more than $17 million.[4] The very earliest of these grants is an excellent example of fundamental research at the nano-bio interface, even illustrating convergence with information science as well.

In 1995, Bruce Schnapp at Harvard University received $349,259 to support "development of a novel optical trapping interferometer for investigations of biological motors." In other words, the aim of this project was to develop a scientific instrument to measure very tiny forces and nanoscale movements inside living cells. In particular, Schnapp wanted to study microtubule motors and vesicle transport, which are absolutely essential to life. Microtubules are a little like structural beams that give the cell its shape and that play important roles in cell division; these structures are about 25 nanometers in diameter but are often several thousand nanometers long, or even a millimeter (million nanometers) in length in long nerve cells. Microtubule motors are nanoscale protein structures, analogous to delivery trucks, that can ride along the microtubule and carry vesicles. Each vesicle is a sort of nanoscale bag, carrying food or waste products from one part of the cell to another.

Twelve years later, Schnapp has a laboratory at Oregon Health and Science University that is dedicated to the study of intracellular transport (i.e., how cargo delivery works) using instruments like those he developed under the grant program:

> Our most recent studies are addressing the questions of how motors are linked to their cargoes and exactly what are these cargoes? These studies are telling us that motors are linked to vesicular cargo via soluble scaffolding proteins that have other functions in the cell, e.g., in scaffolding signal transduction pathways. The idea that motors localize pre-assembled signaling pathways has created a complicated new area for us to explore at the interface of signaling and molecular motors.[5]

Cells are not merely complex chemical factories containing tubules and motor-like proteins: They are also information-processing systems that employ chemical signaling to control their internal processes. Naturally, Schnapp's research instrumentation incorporates computers. Thus, in two very different ways, his research involves information science as well as biology and nanoscience.

The second bio-nano grant, for $300,000 was made in 1996 to Steven Block at Princeton University to develop "an advanced optical trapping microscope facility," a similar kind of research instrumentation. Note that both of the first two nano-bio grants awarded sought to develop the instrumentation necessary to support empirical research in this new area—a very logical stating point. Today, Block has a lab at Stanford University devoted to single-molecule biophysics, or what he likes to call "nature's nanotechnology." Clearly, Block and his team think in terms of convergence:

> Research in our lab marries aspects of physics and biology to study the properties of proteins or nucleic acids at the level of single macromolecules and molecular complexes. Experimental tools include laser-based optical traps ("optical tweezers") and a variety of state-of-the-art fluorescence techniques, in conjunction with custom-built instrumentation for the nanometer-level detection of displacements and piconewton-level detection of forces.[6]

A piconewton is a very tiny amount of force. A newton, named after the famous physicist Sir Isaac Newton, is the force required to accelerate a mass of one kilogram one meter per second during each second it is applied. Coincidentally, this is roughly the same force of gravity that accelerated Newton's famous (but apocryphal) apple when it fell from a tree and inspired his theory of gravitation. Thus a nanonewton is the weight of one billionth of an apple—a "nanoapple," the proverbial speck of dust. A piconewton is 1/1,000 of a nanonewton.[7]

In 1999, a $500,000 grant was made to Marcia Kieliszewski at Ohio University, supporting her research on synthetic plant genes that might be useful in both nanotechnology industries and agriculture. The next year, Corey Smith at Medical College of Georgia received $131,190 to study how certain cells release hormones and neurotransmitter chemicals, with the goal of elucidating principles that could eventually be applied to nanotechnology chemical delivery systems. The next year, four awards totaling $905,000 were made: two installing instrumentation in a lab, one supporting a conference, and the fourth exploring methods for designing protein nanomaterials. This last grant constituted an award of $95,497 to Todd Yeates at UCLA, and was among the first Nanoscale Exploratory Research grants made under the National Nanotechnology Initiative.

In 2002, four grants were issued, two of which were for more than $1 million. A grant for $1,750,000 went to Leonard Rome at UCLA for an NNI

Nanoscale Interdisciplinary Research Team project to explore the nature and possible uses of the cell vesicles called vaults. Shaped somewhat like an American football but only about 60 nanometers long, these structures are called vaults because they look like a collection of nanoscale arches, such as those that form vaulted ceilings in architecture. The hope is that vaults can be used to transport medications inside the cell, to deliver toxins to kill cancer cells, and possibly to remove toxic metals from living tissue. Regardless of whether any of these applications will turn out to be feasible, it is very important to learn what function these distinctive structures fulfill inside our cells. A sense of how convergent this work can be is provided by Rome's remarkable set of affiliations: He is Associate Dean of Research in the UCLA medical school, is Associate Director of the California NanoSystems Institute, is connected with a cancer center plus a brain research institute, and is a biochemistry professor.

The other large grant in 2002 was of a very different kind, consisting of an award of $1,687,733 to Richard O'Grady at the American Institute of Biological Sciences for a series of education and communication activities. Nanotechnology was only incidentally involved in this venture, albeit in a very convergent manner, because a major goal of the project was as follows: "Make the biological community aware of relevant infrastructure and data-networking activities in other scientific/technical communities such as geology, chemistry, toxicology, hydrology, remote sensing/GIS, engineering, nanotechnology, biosensors, mathematics, computer science, and data management/IT."[8]

Neurotransmitter chemicals play a key role in the functioning of the human nervous system, where they act in the synapses between nerve cells. Neurotransmitters are generated in one neuron and are contained in vesicles that may be 50 nanometers in diameter. When released, they flow quickly across the 20-nanometer synaptic gap to receptor sites on the next neuron, where they affect its behavior, thereby contributing to the organism's cognition and action. A 2003 grant for $800,817 to Greg Gerhardt at the University of Kentucky supported development of sensors capable of monitoring neurotransmitter signaling in the brains of living animals. These sensors were small, ranging from 5 to 30 microns rather than being measured in nanometers. Ultimately, the goal will be to shrink at least the key components of the sensors down into the nanoscale.

The formal description of a $100,000 Nanoscale Exploratory Research grant awarded to Eric Furst at the University of Delaware in 2003 gives a very good picture of the revolutionary research many of these "principal investigators" (PIs) are doing:

> One of the major objectives of the National Nanotechnology Initiative is to develop a fundamental understanding of nanoscale

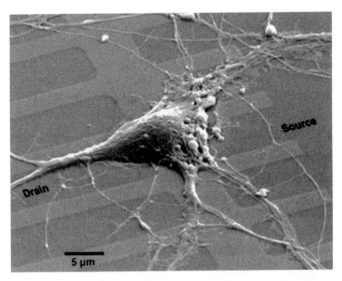

Figure 4–1 A nerve cell from the brain of a rat on a silicon transistor (courtesy Peter Fromherz, Max Planck Institute for Biochemistry). Several groups are doing research at the intersection of all four NBIC fields, in an effort to understand how the brains of living organisms work, and with the ultimate goal of transferring this knowledge to the design of new cognitive and information technologies.

biological structures and processes, such as the mechanics of molecular motor proteins. Molecular motor proteins are enzymes that convert chemical energy directly into mechanical work. These proteins are found in a vast array of biological processes, such as the contraction of smooth and skeletal muscle, cell division, and trafficking materials inside cells and across the cell membrane. Recent experimental advances in motor protein biophysics have resulted in unprecedented insight into the mechanics of single motor molecules. For instance, it was found that the molecular motor myosin makes discrete 5-nm steps, and stalls when opposed by approximately 5 piconewtons (pN) of force. Despite this progress, a major obstacle in motor research has been the inability to resolve the action of the working stroke, when the load-bearing domain of the motor moves forward. The PI proposes to measure the dynamics of the working stroke with sub-nanometer resolution on time scales as small as 25 nanoseconds using a novel light-scattering motor assay.[9]

The largest single grant from this group made in 2006 awarded more than $1 million to Yale University to develop methods for studying the dynamic, complex structures inside cells, thereby enabling researchers to measure forces and movements at many locations simultaneously.[10]

THE PROBLEM OF CANCER

Our new ability to study and manipulate nanoscale phenomena inside the living cell implies that we will make a fresh start in our century-old attempt to conquer that dread disease associated with genetic mutation, cancer. A brief and somewhat bizarre anecdote about my paternal grandfather, William Seaman Bainbridge, will put cancer progress in an historical context. A world-renowned cancer surgeon, in 1914 he published a comprehensive treatise, *The Problem of Cancer*, that was translated into several languages. A century ago, little was known about this disease. Many physicians suspected it was an infectious disease spread by a microbe, such that its development might be prevented by public health measures like those that were conquering typhoid. Dozens of methods of cure were attempted: burning, caustic chemicals, electricity, and the rays of a newly discovered element, radium. A very few excellent surgeons found they could sometimes cure cancer by removing the growth, but the frequent recurrence of the disease taught them they had to perform this excision before metastasis occurred.[11]

Two well-credentialed physicians approached my grandfather, asking his help with their research. Their plan was to feed cancer tissue to birds of prey for several months, to prepare a serum from the birds' blood, and then to inject this serum under the skins of cancer patients. Wild birds of prey ate large numbers of small animals, such as mice, which might carry cancers, yet they were not especially apt to contract the disease themselves. Under the assumption that cancer was an infectious illness, the physicians had decided the birds apparently developed an immunity to it that might be transferred to humans. The two physicians took a place in the country, where they raised birds of prey on human cancers that my grandfather was able to provide them, and then they extracted their serum and injected it into 15 of his patients. At first the patients seemed to improve, but my grandfather quickly discerned that this phenomenon was merely the psychological effect of receiving a hopeful treatment. When none of the 15 actually got better, the experiment was abandoned.[12]

The Problem of Cancer offered much evidence, gathered by many researchers, that cancer was really a collection of similar but distinguishable diseases, but was not primarily infectious in nature. It expressed hope that the right combination of early detection, surgery, and treatments involving chemical agents or radiation might cure many cancers. Over the more than 90 years since my grandfather's book was published, real progress has been achieved in this field, although I tend to think that much of this progress is attributable to modern-day patients' better access to the good care that was available to only a privileged few 50 years ago. If my grandfather were to return to Earth today, I am sure one of his first questions would be whether we had conquered cancer, and I think he would be shocked to learn we have not.

We like to think that medical science is progressing rapidly, but some evidence suggests it has actually passed the point of diminishing returns. The average white American male born in 1900 could expect to live 48 years; this life expectancy had increased to 74 years for a man born in 2000. For white females, the average life expectancy has increased from 51 years to 80 years over the same span. Projecting these figures forward at the same rate of increase suggests that life expectancy in 2100 might be 114 years for males and 125 years for females. Using more realistic assumptions, however, the U.S. Census Bureau has projected that life expectancy for Americans born in the year 2100 will be only 88 years for males and 92 years for females.[13] These Census Bureau estimates are based on the observation that most improvements in longevity came from reducing the risk of death for infants and young adults, whereas it will be more difficult to gain additional years through extending the life span of the elderly.

Progress in health is by no means assured. During the nineteenth century, American health may actually have declined significantly for a number of decades, despite economic growth.[14] We tend to attribute the undeniable improvement in health over the twentieth century to medical progress, but in reality health education and public sanitation may have been more important factors in this trend.[15] The U.S. Centers for Disease Control and Prevention (CDC) argues that substantial improvements in health and longevity could be achieved simply by lifestyle changes—notably, exercising more, reducing the amount of fat in our diet, and avoiding smoking.[16] The introduction of antibiotics made a difference, and modern cardiology saves the lives of many people who might otherwise die prematurely of heart attacks, but the progress in combating cancer has been agonizingly slow. On balance, economists find that the increasing investments in health care are paying off, but not in all areas and not always with very great benefit.[17]

On September 13, 2004, the National Cancer Institute (NCI) announced a five-year, $144 million initiative to apply nanotechnology to cancer. Director Andrew von Eschenbach proclaimed, "Nanotechnology has the potential to radically increase our options for prevention, diagnosis, and treatment of cancer. NCI's commitment to this cancer initiative comes at a critical time. Nanotechnology supports and expands the scientific advances in genomics and proteomics and builds on our understanding of the molecular underpinnings of cancer. These are the pillars which will support progress in cancer."[18]

A booklet published the following January identified five research challenge areas that would be explored, noting that the Human Genome Project had provided vast knowledge related to the genomics and proteomics of human cancers that could now be exploited:[19]

1. Prevention and control of cancer might be achieved by using nanoscale devices to deliver anticancer vaccines or other preventive agents to the areas of a patient's body where genetic analysis suggested a cancer might appear.

2. Implanted nanoscale sensors could detect chemical changes caused by cancer growth, and testing equipment located outside the body could analyze samples for many cancer-associated markers simultaneously.

3. Imaging diagnostics could be improved through the use of "smart" contrast agents that could show doctors exactly which cells were cancerous, and nanoscale devices could map the diversity of cells within a tumor.

4. Diagnosis and treatment could be combined with nanoscale devices that simultaneously deliver therapeutic agents to the cancer while monitoring the results.

5. Nanoscale devices could most effectively deliver medications to improve the quality of life of cancer patients, combating pain, depression, and other secondary symptoms.

A sixth challenge recognizes the convergent nature of nanoscale cancer technology by seeking to develop methods for interdisciplinary training. Under this sort of cross-training, nanotechnology engineers would learn about molecular and systems biology, while cancer researchers would learn about nanoscience. New programs would be established to train researchers, who would then have skills spanning these fields.

In addition, the NCI has announced plans to establish a National Nanotechnology Standardization Laboratory for assessment and standardization of nanodevices. Ideally, the laboratory will accelerate research, development, testing, and deployment of new treatments.

PATHS TO NANO-BIO INNOVATION

Just because nano-bio convergence occurs on a small scale of a few nanometers does not mean that the social organizations creating it will be small. For example, the California NanoSystems Institute was established with funding of $350 million and links the University of California campuses at Los Angeles and Santa Barbara.[20] Three main research thrusts have been established for this institute: (1) nanobiotechnology and biomaterials, (2) nanoelectronics, and (3) nanomechanics and nanofluidics. A somewhat different approach is being tried at the University of Twente in the Netherlands, an institution that styles itself "an entrepreneurial research university." When I visited there in 2005, I toured the substantial Institute for Nanotechnology, which often cooperates with two other "spearhead" laboratories, the Institute for Biomedical Technology and the Center for Telematics and Information Technology.[21] So, one approach is to establish a large institute spanning universities and fields, and another approach is to collect at one university a set of cooperative institutes in the main convergent fields.

The National Nanotechnology Initiative has created networks to connect users with resources across institutions. The National Nanotechnology Infrastructure Network is an "integrated partnership" linking labs at Cornell University, Stanford University, University of Michigan, Georgia Institute of Technology, University of Washington, Pennsylvania State University, University of California at Santa Barbara, University of Minnesota, University of New Mexico, University of Texas at Austin, Harvard University, Howard University, and North Carolina State University.[22] The Network for Computational Nanotechnology links several universities and conducts integrated, cutting-edge research in nanoelectronics, nanoelectromechanical systems, and nano-bioelectronics.[23] Thus another approach is federally supported, distributed networks of facilities.

A very different paradigm connects nano-bio university researchers with small businesses. In 1982, the Small Business Innovation Development Act established a grant funding program across several agencies of the U.S. government (currently a total of 11 agencies): the departments of Agriculture, Commerce, Defense, Education, Energy, Health and Human Services, Homeland Security, and Transportation, plus the Environmental Protection Agency, National Aeronautics and Space Administration, and National Science Foundation. The idea is that economies of scale may allow big businesses to build their own research laboratories, but small businesses may need starter grants if they are to develop new science-based technologies. The agencies designate particular topic areas in which they award Small Business Innovation Research (SBIR) grants, conceptualizing the innovation process in three phases:

- In Phase I, $100,000 start-up grants help a small company explore the potential of a new idea.
- If Phase I is successful, Phase II grants for as much as $750,000 to support research and development move the idea toward commercialization.

- In Phase III, the company exploits the new technology, perhaps attracting investment capital from the marketplace, without government support.[24]

In 2006, NIH announced an SBIR emphasis on "nanotechnologies useful to biomedicine." This agency explicitly identified the need for broad convergence across fields, saying "it is expected that this initiative will require expertise from a variety of disciplines, including engineering, chemistry, physics, material science, engineering, and biology." Recognizing that convergence costs money, to support multidisciplinary teams of researchers, NIH raised the maximum grant size for Phase I to $400,000; for Phase II, the grant size was increased to $1,200,000. While NIH encouraged small companies to suggest fundamentally new ideas, it highlighted eleven priority areas:[25]

- *Nanomaterials (enabling):* Development of synthetic nanoscale building blocks for the formulation of bottom-up approaches to complex and multifunctional nano materials. These materials are expected to find use in applications geared toward pharmaceutical delivery, the development of contrast and biological agents, and multifunctional medical devices.

- *Nano-bio interfaces:* The science of controlling the interface between biomolecular systems and nanoscale synthetic materials, which involves the ability to control the interface architecture and transduction of the control signal through this interface.

- *Nanoimaging:* Real-time imaging of subcellular structure, function, properties, and metabolism.

- *Cell biology:* Nanoscale research on cellular processes, including the biophysics of molecular assemblies, membranes, organelles, and macromolecules.

- *Molecular and cellular sensing/signaling:* Technologies to detect biological signals and single molecules within and outside cells.

- *Prosthetics:* Mechanical, chemical, and cellular implant nanotechnologies to achieve functional replacement tissue architectures and tissue-compatible devices.

- *Environmental and health impact of nanotechnologies:* Ramifications of nanomaterial processing, use, and degradation on health and the environment.

- *In vivo therapeutics:* Development of nanoparticles that enable the controlled release of therapeutic agents, antibodies, genes, and vaccines into targeted cells.

- *Sensor technologies:* Detection and analysis of biologically relevant molecular and physical targets in samples from blood, saliva, and other body fluids, or for use in the research laboratory (purified samples), in clinical specimens, and in the living body.

- *Nanosystem design and application:* Fundamental principles and tools to measure and image the biological processes of health and disease and methods to assemble nanosystems.

- *Bioinformatics for nanotechnology:* Algorithms and computer software to enable and support all of the above.

AGRICULTURE AND THE ENVIRONMENT

World peace and prosperity depend on the existence of an adequate food supply. In a June 2003 report, the U.S. Department of Agriculture (USDA) considered the role that science and technology would play in agriculture in developing countries over the following century, but it failed to mention a very wide range of nanotechnology applications. One somewhat equivocal statement predicted, "Bio-nanotechnology may give molecular biologists even greater opportunities to investigate the physiological functions of plants and animals, which can increase the speed and power of disease diagnosis."[26] Knowledge gained at great cost by nanoscience in rich nations may be applied at far lower cost in poor nations, thus giving nano-enabled research on plant and animal diseases a high priority. Presumably, diagnostic tests based on some form of nanotechnology might be used to benefit farmers in an entire region of a poor country, even if cost concerns limited this testing to only a tiny fraction of their fields and herds.

The USDA report cited far more widespread applications of information technology, including use of the global positioning system (GPS) to guide farm machinery, provide instant information about the market for agricultural products provided over Internet, and assure improved quality control of produce on its way to market. Given the wide range of biotechnologies employed in agriculture, bioinformatics will be crucially important in this arena. In particular, the USDA report devoted much attention to precision agriculture, "a suite of information technologies used to monitor and manage sub-field spatial variability."[27]

The fundamental idea underlying precision agriculture is the use of information technology to determine how much resources and attention crops and animals actually need, and delivery to them of just what they do require. The cost of the information technologies required can best be justified economically when these needs vary greatly across space and time, and when the resources and labor are expensive and, therefore, should not be wasted. For example, irrigation is costly in terms of the water itself, the fuel to run pumps, and the human effort involved in delivering the water. It can be highly useful to know that one field does not need water today, whereas another greatly needs it. In its fully developed form, precision agriculture delivers water exactly where it is needed, to the roots of the plants, minimizing loss through evaporation or runoff. Similarly, pesticides and fertilizers should not be strewn across the landscape, thereby polluting the environment, but rather delivered in the minimum amounts necessary for the crops to be protected and nurtured.

The USDA report explains that a precision farming system today may include the following components:[28]

- Methods for intensively testing soils or plant tissues within a field.
- Equipment for locating a position within a field via GPS.
- A yield monitor.
- A computer to store and manipulate spatial data using some form of geographic information system (GIS) software.
- A variable-rate applicator for seeds, fertilizers, pesticides, or irrigation water.

More-involved systems may also use remote sensing from satellite, aerial, or near-ground imaging platforms during the growing season to detect and treat areas of a field that may be experiencing nutrient stress.

All of this apparatus and attention would be too expensive for small farms and for most farms in poor countries. However, its use on research farms could lead to the development of efficient guidelines for low-tech farmers who are growing the same crops in the same terrain and, therefore, be cost-effective for the information it produces. For high-value crops, in rich countries, and for large farms that can readily amortize the cost of the central information-processing equipment, such a system may already be economical. In addition, some of the information-gathering and information-processing work could be done by governments or on shared computers, thereby reducing the cost for individual farmers.

Much research and development is currently being supported, often in a military context, to develop inexpensive sensor systems that could monitor wide areas, and agricultural applications of this technology can be expected reasonably soon. Such low-cost sensor systems could improve the cost-effectiveness of precision agriculture, not just for plants but also for livestock. The widespread use of radio-frequency identification (RFID) tags is already upon us, and it should be possible in the very near future to monitor the location and condition of every animal on the farm automatically at low cost.

A second USDA report, published only three months after the original report appeared, focused directly on the roles that nanotechnology might play in agriculture, initially in the United States and other advanced nations, and then potentially worldwide. The editors, Norman Scott of Cornell University and Hingda Chen of USDA, argued that nano would chiefly be an enabling technology, usually in combination with information technology:[29]

- Production, processing, and shipment of food products will be made more secure through the development and implementation of nanosensors for pathogen and contaminant detection.

- The development of nanodevices will allow for creation of historical environmental records and location tracking of individual shipments.

- Systems that provide the integration of "smart system" sensing, localization, reporting, and remote control will increase efficiency and security.

- Agricultural and food systems security is of critical importance to homeland security. Our nation's food supply must be carefully monitored and protected. Nanotechnology holds the potential of such a system becoming a reality.

The U.S. Environmental Protection Agency (EPA) showed an early interest in nanotechnology, and by October 2006 it had posted six reports and workshop proceedings on its nano website.[30] On August 28–29, 2002, EPA held a workshop for researchers who had received nanotechnology grants.[31] Most of these projects fell into one of three areas: (1) sensors, (2) treatment, or (3) remediation. Four projects were developing nano-enabled sensors for testing water quality, and one was developing new methods to identify airborne nanoscale particles. In the treatment category, researchers were exploring the value of nanoscale biopolymers for removing dangerous heavy metals, nano-improved catalytic converters to reduce pollution emissions from vehicles, nanostructured materials for solar power collection systems, the potential of nanocrystalline materials as catalysts for emission abatement, and fundamental research on nanoparticle production. Remediation research explored ways of using nanotechnology to remove hazardous materials from contaminated sites.

An EPA workshop held October 20–21, 2005, was devoted entirely to remediation, a huge problem—"there are between 235,000 and 355,000 sites in the country requiring cleanup at an estimated cost of between [$174 billion and $253 billion]."[32] Participants noted, for example, that the standard way of dealing with contaminated groundwater is to pump it out of the ground for treatment, yet new methods such as the use of nanoscale iron particles [zero valent iron (ZVI)] might make it possible to treat the groundwater in situ.

On December 2, 2005, EPA offered a draft nanotechnology white paper online for comment, prior to considering whether to give this document official status. Based on deliberations among about five dozen experts, the white paper sought to assess the balance between nanotechnology's potential dangers to health and the environment against nanotechnology's ability to help protect humanity from hazards. The group classified intentionally produced nanomaterials into four groups: (1) carbon-based materials, such as buckyballs (spherical structures of carbon atoms) and nanotubes (cylindrical structures); (2) metal-based materials, such as quantum dots, which can be highly reactive because the majority of their atoms are on their surfaces; (3) dendrimers (root-like structures) composed of nanoscale branched polymers; and (4) composites consisting of two or more kinds of nanoparticles or in which nanoparticles are embedded in a bulk material. All of these nanomaterials are expected to play important roles in future industrial products, and research is needed to characterize any potential hazards and find the best responses to them. The white paper explicitly predicted convergence:

> In the long term, nanotechnology increasingly will likely be discussed within the context of the convergence, integration, and synergy of nanotechnology, biotechnology, information technology, and cognitive technology. Convergence involves the development of novel products with enhanced capabilities that incorporate bottom-up assembly of miniature components with accompanying biological, computational, and cognitive capabilities. The convergence of nanotechnology and biotechnology, already rapidly progressing, will result in the production of novel nanoscale materials. The convergence of nanotechnology and biotechnology with information technology and cognitive science is expected to rapidly accelerate in the coming decades. The increased understanding of biological systems will provide valuable information towards the development of efficient and versatile biomimetic tools, systems, and architecture.[33]

EVOLUTIONARY METHODS: COMPUTING AND CULTURE

The most fundamental concept of biology is evolution through natural selection, based on mutation and inheritance of genetic codes embodied in the DNA molecule. Unfortunately, many Americans reject this fact or at least are uncomfortably uncertain about it. One study comparing 34 nations found that the 40 percent of Americans who accept the theory of evolution is smaller than the percentage of people who accept this theory in all 32 European nations in the list but higher than the percentage in the only Islamic nation studied, Turkey.[34] In contrast, at least 80 percent of citizens are sure the theory is true in Denmark, France, Iceland, and Sweden; 78 percent of the Japanese population accepts evolutionary theory.

In the field of medicine, we see the constant working of evolution in the emergence of drug-resistant strains of bacteria and the transformation of viral diseases such as influenza and AIDS as they transfer from animals to humans. The Human Genome Project and similar efforts to sequence the human genetic code are preparing the way for new forms of diagnosis and treatment tailored to the individual genetic makeup of the patient, incidentally documenting for all to see the fundamental facts of genetic variation and transmission.

Gene sequencing works at the nanoscale, because the diameter of the thread-like DNA molecule is about 2.5 nanometers. However, the convergent technology that was absolutely necessary for sequencing the human genome was information technology. The *shotgun method* of sequencing breaks many copies of a given long DNA sample into fragments that can then be analyzed. In the original article that announced substantial completion of the Human Genome Project, one hypothetical sequence used for illustration was ACCG-TAAATGGGCTGATCATGCTTAAA.[35] The letters represent the base pairs (adenine, cytosine, guanine, and thymine) that define the code. Another DNA fragment might have the sequence TGATCATGCTTAAACCCTGTG-CATCCTACTG. A computer searching thousands of fragments for common sequences would find that these two fragments shared the TGATCATGCT-TAAA sequence and combine them to get a longer sequence, ACCG-TAAATGGGCTGATCATGCTTAAACCCTGTGCATCCTACTG. Step by step, using efficient algorithms developed by computer scientists, the information technology can then assemble the entire chromosome's code.

The initial sequencing of the human genome is merely the beginning of the research challenge, however. In truth, we have just begun to understand human genetic variation, including variable disease resistance and susceptibility to drug side effects, and the translation of the genetic code into the production of proteins and determination of human characteristics. Proteomics studies will most assuredly require nano-bio research to determine how proteins function, and information science projects to model their structure, dynamics, and interactions.

A multidisciplinary, four-institution team led by computer scientist Raj Reddy at Carnegie Mellon University received more than $8 million in NSF grants to see what could be achieved if we take seriously the metaphor that

DNA carries a genetic "code" analogous to codes in language.[36] Reddy's team applied conceptualizations and techniques from computational linguistics to genetic data. Both fields start with data—raw text in databases or libraries in the case of language, and large numbers of whole or fragmentary genome sequences in the case of genetics. Both then map these data to larger functional structures. That is, gene sequences produce proteins with particular configurations, functions, and forms of activity. Sequences of letters of the alphabet gives rise to words, which subsequently take on meanings in sentences and paragraphs. Both results must be decoded by computers and human minds to be understood through processes of information retrieval, summarization, and translation.

An example is the technique called *n-grams* used in *natural language processing* (NLP), which is a general term for computer analysis of written or spoken language.[37] An *n*-gram is a statistical or probabilistic method of analysis used in a variety of research strategies and applications, such as dialog systems that allow a human to talk with a computer (with at least some degree of success). Words often have multiple meanings, which both people and computers must attempt to disambiguate. One of the best ways to tell which meaning of a word is intended is to look at other words in the sentence and paragraph.

Some NLP approaches look at a paragraph as a "bag of words" and do not worry about their precise order. By contrast, *n*-grams specifically look at strings of words in order. A 1-gram is a single word, such as "converging." A 2-gram is a pair of words, such as "converging technologies." A 3-gram is a triplet, such as "converging technologies for." A 4-gram is a quartet, such as "converging technologies for improving." An *n*-gram is the abstract idea of a string of *n* words. If you see one word, it is seldom possible to predict what the next word in the sentence might be. With two words, however, you have a better chance of predicting the third, and so on.

Using Google on September 24, 2006, I experimented with the phrase "converging technologies for improving human performance." When I entered just "converging" into the search engine, it found 8,450,000 webpages using the word. By comparison, "converging technologies" turned up just 243,000 pages. "Converging technologies for" produced 45,900 pages. Adding the next three words of the phrase in order gave 28,700, 27,900, and 26,200 pages. Notice that once I had added the word "improving," the entire phrase was pretty much determined, and the number of pages ceased dropping so rapidly.

Reddy and his research team suggested that application of *n*-grams and other NLP methods could help identify errors in genome sequences and fill in puzzling gaps. These methods work best when vast amounts of data are available for analysis. For example, to predict the next word in a sequence, we would first analyze a huge body of naturally occurring text to see how often that word is ever used. Next, computers would count how many times various other words are used immediately before and after it, and so on. In the case of genomics, we would need vast stores of genetic data not only about individual humans but also about many other species related to humans in varying degrees. Thus computational approaches to genomics exemplify a concept mentioned in Chapter 1—the new style of "industrial, mass-production science," in which armies of technicians and fleets of computers collect and analyze vast amounts of data describing very complex systems.

For decades, innovative scientists have been suggesting the opposite of what I have just described. That is, they have sought to apply concepts from biology to other fields such as computers and social science, notably the principle of evolution by natural selection. Clearly, converging fields such as biology and computer science can make mutual contributions to each other, rather than one being merely a tool to accomplish the goals of the other.

Especially noteworthy is the general approach to computing called *genetic algorithms* or *evolutionary computing*. Perhaps the first full statement of this idea appeared in a 1975 book, *Adaptation in Natural and Artificial Systems* by John Holland.[38] In genetic algorithms, one writes a program such that a string of letters comparable to the DNA code represents a possible solution to a complex problem. For example, a sequence of L's, R's, and F's could represent turning left, turning right, or going forward in a maze, respectively. At its start, the program creates a very large number of such strings, with every letter being chosen at random. The program then tests each string and assigns it a score based on how well or how poorly it solves the problem—that is, how closely the string matches the maze. Next, a new generation of strings is created, in which several copies are made of the strings that got the highest scores, and few if any copies are made of the strings with poor scores.

The program can be set to create an error during a certain fraction of the time—that is, inserting one letter at random instead of copying one from the parent string. This is equivalent to mutation in biological evolution, when an incorrect base pair is inserted by mistake. Most mutations are harmful, but a very few are beneficial. Evolution progresses most quickly if beneficial mutations and beneficial combinations of genes can be combined. Therefore, vertebrate animals have male and female sexes, which allow for random combination of genes across lineages. In genetic algorithms, the term *crossover* is used to describe this phenomenon rather than *intercourse*, probably to preserve the dignity of computer science. Some fraction of the second-generation strings will have a combination of the codes of two parents. This second generation is tested, and a third generation is then created that increases the proportion of high-scoring strings. Given enough generations, a good scoring system, and a solvable problem, the problem will, indeed, be solved.

At the first Converging Technologies conference, Jordan Pollack from Brandeis University reported on his success in designing robots by evolutionary methods. When Pollack's research was picked up by the popular press, it was distorted. As a consequence, some otherwise reasonable people got the false impression he was actually breeding robots the way people breed horses, or the way that mythical self-reproducing nanobots might produce offspring. Rather, Pollack was running evolutionary programs inside a computer, letting different designs compete within a virtual world. Graduate students then used hand tools to build the actual robots out of components such as rods and pistons, on the basis of the computer-generated designs. This was enough of an accomplishment, and robots evolving in this way were, in fact, able to scamper faster and faster across the floor with each successive generation (Figure 4–2).

Pollack was not merely playing with toys or using the robots produced in this way to demonstrate the fruitfulness of genetic algorithms. Rather, he was developing an approach to tackle the grand scientific problem of our age: how to understand, control, and build systems of vast complexity. Pollack had

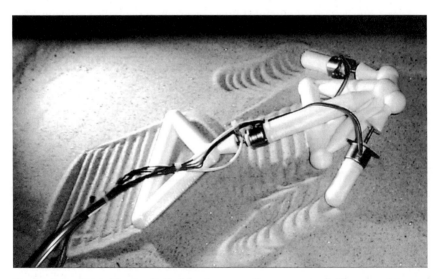

Figure 4–2 One of Jordan Pollack's robots crawling across sand. These machines were designed following the principles of evolution by natural selection from random variation, showing how a fundamental concept from biology can be applied to engineering design completely outside the medical area.

recognized the very limited nature of progress in artificial intelligence, the large number of bugs that wind up in big computer programs coded by people, and the failure of computer science to develop computing languages that would solve these problems. He argues that the solution is bio-info convergence:

> The opportunity available today is that the way out of this tarpit, the path to achieving both software and nano devices of biological complexity with tens of billions of moving parts, is very clear: It is through increasing our scientific understanding of the processes by which biologically complex objects arose. As we understand these processes, we will be able to replicate them in software and electronics. The principles of automatic design and of self-organizing systems are a grand challenge to unravel. Fortunately, remarkable progress has been shown since the computer has been available to refine the theory of evolution. Software is being used to model life itself, which has been best defined as that "chemical reaction, far from equilibrium, which dissipates energy and locally reverses entropy."
>
> Much as logic was unconstrained philosophy before computer automation, and as psychological and linguistic theories that could not be computerized were outgunned by formalizable models, theories on the origin of life, its intrinsic metabolic and gene regulation processes, and the mechanisms underlying major transitions in evolution are being sharpened and refuted through formalization and detailed computer simulation.[39]

If convergence between computing and biology has been going on for three decades, the same is true for convergence between social science and biology. Edward O. Wilson published his tremendously influential book *Sociobiology: The New Synthesis* in 1975, coincidentally the same year that Holland's less well-known but also influential work appeared.[40] Wilson, of course, is the visionary who more recently promoted the idea that the sciences are uniting in consilience. His sociobiology book sought to synthesize a new multidisciplinary field in which animal and human behavior was understood to be the result of biological evolution. One year later, in *The Selfish Gene*, Richard Dawkins argued that culture evolves just as living organisms do; he proposed *meme* as the name for the cultural equivalent of gene, a unit of human imitation.[41] In 1981, with Charles Lumsden, Wilson published *Genes, Mind, and Culture*, applying evolutionary ideas even more fully to culture.[42] In that same year, Luigi Luca Cavalli-Sforza and Marcus W. Feldman published a parallel statement, *Cultural Transmission and Evolution*.[43]

Three things are necessary to support a cultural genetics:

1. There must be some process of reproduction and inheritance, in which cultural structures and elements are transmitted from one "generation" to the next.

2. There must be a significant measure of stability in the transmission process, in which the replicators show sufficient copying fidelity to transmit recognizable patterns.

3. There must be some process such as sexuality or mutation that introduces change and variety into the process of inheritance, yet is sufficiently coherent itself to permit scientific analysis.

At a conference in 1982, I argued that these conditions are met by religious cults and by at least some other phenomena such as stylistic schools in the various arts.[44] If other parts of the wider culture fail to exhibit these features, an "inorganic chemistry" of culture—if not the full richness of an organic genetics—will still emerge, and the rules of one can illuminate the rules of the other.

Cult is culture writ small.[45] When I proposed this aphorism in 1978, I meant to suggest that cults are the *Drosophila melanogaster* and *Escherichia coli* that will permit us to develop cultural genetics—in other words, fast-reproducing organisms whose evolution can be studied efficiently. In modern society, cults are born out of older cults through two processes biologists call fission and sporulation, and most cults are known to cluster in family lineages. *Fission* is the common term for reproduction by splitting among microorganisms; the corresponding term in religious studies is *schism*. *Sporulation* is the term used in biology to name the process by which organisms of certain species (the mosses, for example) reproduce by throwing off *spores* (tiny seeds). Similarly, the founders of religious cults almost invariably serve an apprenticeship in earlier successful cults, so they serve as the seeds in this reproduction process. In this way, the beliefs and behaviors of cults and artistic movements are transmitted from one generation to the next, with these characteristics sometimes mutating and often combining from two or more sources.

At the first Converging Technologies conference, my NSF colleague Gary Strong and I argued in favor of the development of a new social science, which we called *memetics*, following an evolutionary approach to the analysis of culture.[46] We noted then that the most valuable resource in the information society of the twenty-first century will be not iron or oil, but rather culture. An evolutionary science of culture would need to begin with classification studies—delineating, for example, the species and genera of music—but could soon be used to create new cultural objects by combining elements of existing cultural objects, by employing innovative mutation-generating mechanisms, and by observing which areas of cultural territory are not already occupied. By *culture*, we mean not just art, music, language, clothing styles, and ethnic foods, but also technology, which already is being designed occasionally using evolutionary computing.[47] In addition, we include the fundamental values, norms, and beliefs that define a society's way of life as part of our definition of culture.

Clearly, memetics could become controversial as well as powerful if it began to make real progress as a convergent science and technology. To accomplish this transformation, Strong and I argued, three things would be needed:

- Professional conferences, scientific journals, and a formal organization devoted to memetics.

- Data infrastructure, in the form of multiuse, multiuser digital libraries incorporating systematic data about cultural variation, along with software tools for conducting scientific research on it.

- Specific major research projects assembling multidisciplinary teams to study distinct cultural phenomena that are most likely to advance fundamental memetic science and to have substantial benefits for human beings.

Although the idea of applying evolutionary concepts to the study and design of culture is not new, only since the development of the World Wide Web in the 1990s have computers had access to vast enough stores of information about culture first to develop a comprehensive taxonomy, and then to begin serious genetic engineering of the worlds of art, inventions, and other cultural innovations.[48] At both the first and second Converging Technologies conferences, Robert Horn urged undertaking a "Human Cognome Initiative" to map the human mind in much the same way the Human Genome Project had mapped our DNA.[49]

In my view, while this research would include a complete mapping of the connections in the human brain, its scope would be far more extensive than neuroscience. The archaeological record indicates that anatomically modern humans existed tens of thousands of years before the earliest examples of art appeared, a fact that suggests the human mind was not merely the result of brain evolution but also required substantial evolution in culture and personality. Central to the Human Cognome Initiative would be wholly new kinds of rigorous research on the nature of both culture and personality, in addition to fundamental advances in cognitive science.

IMPROVING HUMAN PERFORMANCE

Consider the full titles of the first two NBIC reports: *Converging Technologies for Improving Human Performance* and *The Coevolution of Human Potential and Converging Technologies*. What do the phrases "human performance" and "human potential" mean in this context?

One view is that technological convergence will give humanity unprecedented powers to shape our environment. The very title of the most recent NSET report, *Nanotechnology: Societal Implications, Maximizing Benefits for Humanity*, suggests nanoconvergence can do wonderful things for humanity. One task force contributing to that report was asked to examine NBIC, and it reached the following conclusion: "Over the coming decades, the world may be transformed by the convergence of four major realms of discovery and invention: nanotechnology, biotechnology, information technology, and new technologies based in the cognitive sciences."[50] As a part of the world, will the human species also be transformed?

Many participants in the nanoconvergence conferences are uncomfortable with the idea of human transformation, both because they find the prospect personally uncomfortable and because they think the idea is politically dangerous. Setting extremely revolutionary goals for nanoconvergence is a good way to lose support from the general public, they think. Conversely, other participants believe that many of the health and welfare benefits to be gained from nanoconvergence will come through genetic engineering (i.e., preventive biotechnology treatments that incidentally change the human body) and through cognitive enhancements achieved by the convergence of all four NBIC fields. A significant minority among these nano proponents believe that the transformation of human nature is a good thing in itself, effectively taking biological evolution into our own hands to create descendants who are not only healthier and happier, but also smarter and morally superior.

At the risk of oversimplification, convergence is the third of three great technological revolutions in human history, in which new abilities to transform the world also transformed humanity itself. The first was the Neolithic Revolution, which occurred thousands of years ago, when the development of agriculture permitted the emergence of cities, a huge increase in human population, and a proliferation of new skills and technologies such as writing and the beginning of chemistry as represented by the use of metals. The scholar who did the most to clarify the importance of the Neolithic Revolution, V. Gordon Childe, called his 1951 book *Man Makes Himself*, announcing that we ourselves were transformed in the process of transforming the world.[51]

The second revolution was, of course, the Industrial Revolution, which began in England approximately 250 years ago and culminated in the mass-production methods perfected in the United States in the early twentieth century. Although the early years of this revolution may have been difficult for the working classes, and industrial society rewards capitalists more than creators, the Industrial Revolution had democratizing characteristics, in that economic growth and communication technologies supported universal literacy and suffrage. People's bodies grew taller, fatter, and less afflicted by parasites and infectious diseases. After learning to read, they learned to drive cars and developed a host of technology-oriented skills their ancestors lacked.

Clearly, technological advances of the past have often transformed humans in some significant degree, even as they were changing the conditions of life. Now, however, biotechnology supported by the other fields appears poised to change human nature in fundamental ways. Chapter 5 will consider cognitive enhancements at length, so here we concentrate on physical enhancements and the more general ethical issues.

In the first NBIC report, a scientific task force reported on the likely future methods for improving human health and physical capabilities.[52] The members of the task force highlighted six representative examples:

1. Research on disease and nutrition could be greatly advanced through a nano-bio processor—that is, a nano-bio-info device that programmed biological processes on a chip where they could be studied simultaneously at the level of tiny molecular details and at the level of entire systems.

2. Nano-enabled implant devices could monitor a person's physiological well-being in real time, warning of impending diseases and other conditions that could be handled by early interventions.

3. Without subscribing to the idea of molecular nanobots, it should be possible to design tiny robots that employ nanoscale components to carry out microsurgery and other tasks inside the human body.

4. Sensorily disabled people—especially those with impaired vision and hearing—could benefit from multimodal information communication devices that use to the maximum extent possible the senses that the individual does possess.

5. Brain-to-brain and brain-to-machine interfaces would first be valuable research tools, then lead to early applications for disabled individuals and in military contexts, and eventually enhance the lives of ordinary people.

6. Virtual environments, exploiting advanced knowledge of human senses and the brain, would be valuable in education, design and evaluation of major architectural projects, scientific or workgroup collaboratories, and entertainment.

Patricia Connolly of the University of Strathclyde argued that convergence could extend the life span.[53] Michael Heller of the University of California at San Diego suggested that we need to understand better the connection between genetics and performance in such areas as sports and mental health, and Jeffrey Bonadio of the University of Washington surveyed the potential of gene therapy to improve health.[54] Peter C. Johnson, representing Tissue Informatics Corporation, explained the essential role that information technology can play in supporting medical research and the development of new biotechnology through bioinformatics.[55]

From the very first NBIC conference, it was plain that the drive to improve human performance raised serious questions about who should apply standards regarding what is considered normal, desirable, and ethical. One voice was heard very clearly, that of Gregor Wolbring of the University of Calgary.[56] His current blog describes him and his perspective:

> I am a social Entrepreneur a social-Preneur. I am a thalidomider and a wheelchair user. I am a biochemist and a bioethicist. I am a scientist and an activist. I work on issues related to bioethics, health research, disabled and other marginalized people's and human rights, governance of science and technology, and evaluation of new and emerging technologies. I am the founder of the International Centre for Bioethics, Culture and Disability and of the International Network on Bioethics and Disability. I believe that a wide open public debate on how the above issues affect society and marginalized groups is the only way to develop safeguards against abuse.[57]

In an age when technological convergence renders many previously impossible things possible, Wolbring suggests that the question no longer concerns how normal people can help the disabled, but rather how differently abled people can help themselves. The issue is not how "normal" people can compensate "handicapped" people for their deficiencies; it is how each person, whatever his or her particular mix of abilities, will be able to grow in capabilities, to thrive, and to develop in whatever direction this individual sees fit. This approach requires a more socially enlightened viewpoint, rather than the medical view that every unusual condition is a disease requiring a cure.

On a deep level, this viewpoint is compatible with the view that "normal" people deserve the right to use biotechnology to enhance their native abilities. In the fourth NBIC book-length report, *Progress in Convergence*, Andy Miah notes the necessity of rethinking enhancement in sport.[58] The boundary between permissible medical treatment for an athlete's illnesses and unfair enhancements is unclear. Miah notes the growing use of hypoxic training, in which an athlete exercises in low-oxygen or low-pressure chambers, thereby causing the body to adapt by producing more red blood cells and other changes that give the athlete greater endurance in competitions under normal atmospheric conditions. If this practice is unfair, then what about allowing athletes to train in high-altitude locations, such as La Paz, Bolivia? Why is there an ethical difference? The same book also contains a parallel analysis by Oxford University philosopher Julian Savulescu, establishing the ethical basis for transcendence of the biological human condition.

CONCLUSION

We should not be afraid of change, nor should we be afraid of well-considered actions to steer the direction of change. Biotechnology already has the power to transform human life in many ways—for example, through genetics, a field fraught with controversy. Convergence with the three other NBIC fields increases the power of biotechnology, but it also broadens our range of options and introduces new scientific tools with which we can assess the likely outcomes of our decisions. The NBIC task force that authored the most recent major report on the scientific implications of nanotechnology advocated a wise and balanced approach to this issue:

- Technological convergence has the potential to achieve benefit for all human beings, if it does not become ensnared in unnecessary complexity, uncertainty, and public alienation. Therefore, actions such as the following should be undertaken:

- Research and education should be promoted about best practices in organizational design, so that organizations such as corporations and universities will have greater flexibility to change and be faster in making beneficial changes.

- To achieve a regulatory environment that protects the public while encouraging innovations, there should be greater coordination and consistency across federal regulatory agencies, expanded coordination between federal and state regulators, and efforts to establish common regulatory frameworks with other nations.

■ Federal agencies should engage in research to understand, quantify, and mitigate risks to human health and well-being that arise from convergence of NBIC technologies.

■ Government should experiment with new mechanisms for agencies to engage the public to provide a citizens' perspective.

■ Global frameworks should be considered that offer common parameters for research, harmonization of regulations, and market access that could expedite the development and commercialization of beneficial NBIC technologies by reducing risk, creating transparency, and contributing to a level playing field for all competitors.[59]

Note that these recommendations concern the social structures and deliberative contexts through which the world will make decisions, not the content of the decisions themselves. This implies that the cognitive and information sciences will be crucial for developing effective means for developing policy about the convergence of nanotechnology and biotechnology.

REFERENCES

1. John H. Marburger III, "The Future of Nanotechnology," in Mihail C. Roco and William Sims Bainbridge (eds.), *Nanotechnology: Societal Implications—Maximizing Benefit for Humanity* (Arlington, VA: National Nanotechnology Coordination Office, 2006, p. 13), online at http://www.nano.gov/nni_societal_implications.pdf

2. Viola Vogel and Barbara Baird (eds.), *Nanobiotechnology* (Arlington, VA: National Nanotechnology Coordination Office, 2005, p. 4).

3. National Science Foundation, *Grant Proposal Guide* (Arlington, VA: National Science Foundation, 2004, p. 10), online at http://www.nsf.gov/pubs/gpg/nsf04_23/nsf04_23.pdf

4. http://www.nsf.gov/awardsearch/index.jsp

5. http://www.ohsu.edu/cellbio/faculty/faculty%20pages/schnapp_pages/index.html

6. http://www.stanford.edu/group/blocklab/ResearchMain.htm

7. Mark C. Williams, "Optical Tweezers: Measuring Piconewton Forces," 2004, online at http://www.biophysics.org/education/williams.pdf

8. http://www.nsf.gov/awardsearch/showAward.do?AwardNumber=0229195

9. http://www.nsf.gov/awardsearch/showAward.do?AwardNumber= 0304051

10. http://www.nsf.gov/awardsearch/showAward.do?AwardNumber= 0619674

11. L. J. Rather, *The Genesis of Cancer: A Study in the History of Ideas* (Baltimore, MD: Johns Hopkins University Press, 1978); James T. Patterson, *The Dread Disease: Cancer and Modern American Culture* (Cambridge, MA: Harvard University Press, 1987).

12. William Seaman Bainbridge, *The Cancer Problem* (New York: Macmillan, 1914, p. 251).

13. F. W. Hollmann, T. J. Mulder, and J. E. Kallan, *Methodology and Assumptions for the Population Projections of the United States: 1999 to 2100*, Population Division Working Paper No. 38 (Washington, DC: U.S. Census Bureau, 2000).

14. D. L. Costa and R. H. Steckel, "Long-Term Trends in Health, Welfare, and Economic Growth in the United States," in R. H. Steckel and R. Floud (eds.), *Health and Welfare During Industrialization* (Chicago: University of Chicago Press, 1997, pp. 47–89).

15. Samuel H. Preston, *American Longevity: Past, Present, and Future* (Syracuse, New York: Center for Policy Research, Syracuse University, 1996).

16. Centers for Disease Control and Prevention, U.S. Department of Health and Human Services, *Chronic Diseases and Their Risk Factors: The Nation's Leading Causes of Death* (Atlanta, GA: Centers for Disease Control and Prevention, 1999); *Reducing Tobacco Use: A Report of the Surgeon General* (Atlanta, GA: Centers for Disease Control and Prevention, 2000).

17. D. M. Cutler, M. McClellan, J. P. Newhouse, and D. Remler, "Are Medical Prices Declining?" *Quarterly Journal of Economics*, 113:991–1024, 1998; D. M. Cutler, and E. Richardson, "Your Money and Your Life: The Value of Health Care and What Affects It," in A. M. Garber (ed.), *Frontiers in Health Policy Research* (Cambridge, MA: MIT Press, 1999, pp. 99–132); V. R. Fuchs, "The Future of Health Economics," *Journal of Health Economics*, 19:141–157, 2000.

18. NCI Press Office, "National Cancer Institute Announces Major Commitment to Nanotechnology for Cancer Research," *NIH News,* September 13, 2004, online at http://www.nih.gov/news/pr/sep2004/nci-13.htm

19. National Cancer Institute, *Cancer Nanotechnology: Going Small for Big Advances* (Bethesda, MD: National Cancer Institute, 2004), online at http://otir.nci.nih.gov/brochure.pdf

20. http://www.cnsi.ucla.edu/staticpages/about-us

21. http://www.utwente.nl/en/matrix/institutes/bsmpae-spearhead/

22. http://www.nnin.org/

23. http://www.ncn.purdue.edu/about/research/

24. http://www.sba.gov/SBIR/indexsbir-sttr.html

25. http://grants2.nih.gov/grants/guide/pa-files/PA-06-009.html

26. U.S. Department of Agriculture, "21st Century Agriculture: A Critical Role for Science and Technology," June 2003, p. 24, online at http://www.usda.gov/news/pdf/agst21stcentury.pdf

27. U.S. Department of Agriculture, "21st Century Agriculture: A Critical Role for Science and Technology," June 2003, p. 19, online at http://www.usda.gov/news/pdf/agst21stcentury.pdf

28. U.S. Department of Agriculture, "21st Century Agriculture: A Critical Role for Science and Technology," June 2003, p. 18, online at http://www.usda.gov/news/pdf/agst21stcentury.pdf

29. Norman Scott and Hongda Chen (eds.), *Nanoscale Science and Engineering for Agriculture and Food Systems* (Ithaca, NY: Cornell University, 2003, p. 15), online at http://www.nseafs.cornell.edu/roadmap.draft.pdf

30. http://es.epa.gov/ncer/nano/publications/index.html

31. http://es.epa.gov/ncer/nano/publications/2002_august_nano_star_workshop.pdf

32. U.S. EPA Workshop on Nanotechnology for Site Remediation, 2005, p. 3, online at http://es.epa.gov/ncer/publications/workshop/pdf/10_20_05_nanosummary.pdf

33. Jeff Morris and Jim Willis (eds.), *Nanotechnology White Paper* (External Review Draft, U.S. Environmental Protection Agency), 2005, p. 7, online at http://es.epa.gov/ncer/nano/publications/whitepaper12022005.pdf

34. Jon D. Miller, Eugene C. Scott, and Shinji Okamoto, "Public Acceptance of Evolution," *Science,* 313:765–766, 2006.

35. International Human Genome Sequencing Consortium, "Initial Sequencing and Analysis of the Human Genome," *Nature,* 409:860–921, 2001.

36. Judith Klein-Seetharaman and Raj Reddy, "Biological Language Modeling: Convergence of Computational Linguistics and Biological Chemistry," in Mihail C. Roco and William Sims Bainbridge (eds.), *Converging Technologies for Improving Human Performance* (Dordrecht, Netherlands: Kluwer, 2003, pp. 428–437).

37. James Martin, "Natural Language Processing," in William Sims Bainbridge (ed.), *Encyclopedia of Human–Computer Interaction* (Great Barrington, MA: Berkshire, 2004, pp. 495–501).

38. John H. Holland, *Adaptation in Natural and Artificial Systems* (Ann Arbor, MI: University of Michigan Press, 1975).

39. Jordan Pollack, "Breaking the Limits on Design Complexity," in Mihail C. Roco and William Sims Bainbridge (eds.), *Converging Technologies for Improving Human Performance* (Dordrecht, Netherlands: Kluwer, 2003, p. 162).

40. Edward O. Wilson, *Sociobiology: The New Synthesis* (Cambridge, MA: Belknap Press of Harvard University Press, 1975).

41. Richard Dawkins, *The Selfish Gene* (New York: Oxford University Press, 1976).

42. Charles J. Lumsden and Edward O. Wilson, *Genes, Mind, and Culture* (Cambridge, MA: Harvard University Press, 1981).

43. Luigi Luca Cavalli-Sforza and Marcus W. Feldman, *Cultural Transmission and Evolution* (Princeton, NJ: Princeton University Press, 1981).

44. William Sims Bainbridge, "Cultural Genetics," in Rodney Stark (ed.), *Religious Movements* (New York: Paragon, 1985, pp. 157–198).

45. William Sims Bainbridge, *Satan's Power: A Deviant Psychotherapy Cult* (Berkeley, CA: University of California Press, 1978, p. 14).

46. Gary W. Strong and William Sims Bainbridge, "Memetics: A Potential New Science," in Mihail C. Roco and William Sims Bainbridge (eds.), *Converging Technologies for Improving Human Performance* (Dordrecht, Netherlands: Kluwer, 2003, pp. 318–325).

47. William Sims Bainbridge, "Evolutionary Engineering," in William Sims Bainbridge (ed.), *Encyclopedia of Human–Computer Interaction* (Great Barrington, MA: Berkshire, 2004, pp. 244–247).

48. William Sims Bainbridge, "The Evolution of Semantic Systems," in Mihail C. Roco and Carlo D. Montemagno (eds.), *The Coevolution of Human Potential and Converging Technologies* (New York: New York Academy of Sciences, 2004, pp. 150–177).

49. Robert E. Horn, "Visual Language and Converging Technologies in the Next 10–15 Years (and Beyond)," in Mihail C. Roco and William Sims Bainbridge (eds.), *Converging Technologies for Improving Human Performance* (Dordrecht, Netherlands: Kluwer, 2003, pp. 141–149); "To Think Bigger Thoughts: Why the Human Cognome Project Requires Visual Language Tools to Address Social Messes," in Mihail C. Roco and Carlo

D. Montemagno (eds.), *The Coevolution of Human Potential and Converging Technologies* (New York: New York Academy of Sciences, 2004, pp. 212–220).

50. Mihail C. Roco and William Sims Bainbridge (eds.), *Nanotechnology: Societal Implications, Maximizing Benefits for Humanity* (Arlington, VA: National Nanotechnology Coordination Office, 2005, p. 59).

51. V. Gordon Childe, *Man Makes Himself* (New York: New American Library, 1951).

52. Mihail C. Roco and William Sims Bainbridge (eds.), *Converging Technologies for Improving Human Performance* (Dordrecht, Netherlands: Kluwer, 2003, pp. 179–182).

53. Patricia Connolly, "Nanobiotechnology and Life Extension," in Mihail C. Roco and William Sims Bainbridge (eds.), *Converging Technologies for Improving Human Performance* (Dordrecht, Netherlands: Kluwer, 2003, pp. 182–190).

54. Michael J. Heller, "The Nano-Bio Connection and Its Implication for Human Performance," in Mihail C. Roco and William Sims Bainbridge (eds.), *Converging Technologies for Improving Human Performance* (Dordrecht, Netherlands: Kluwer, 2003, pp. 191–193); Jeffrey Bonadio, "Gene Therapy: Reinventing the Wheel or Useful Adjunct to Existing Paradigms?" in Mihail C. Roco and William Sims Bainbridge (eds.), *Converging Technologies for Improving Human Performance* (Dordrecht, Netherlands: Kluwer, 2003, pp. 194–207).

55. Peter C. Johnson, "Implications of the Continuum of Bioinformatics," in Mihail C. Roco and William Sims Bainbridge (eds.), *Converging Technologies for Improving Human Performance* (Dordrecht, Netherlands: Kluwer, 2003, pp. 207–213).

56. Gregor Wolbring, "Science and Technology and the Triple D (Disease, Disability, Defect)," in Mihail C. Roco and William Sims Bainbridge (eds.), *Converging Technologies for Improving Human Performance* (Dordrecht, Netherlands: Kluwer, 2003, pp. 232–243).

57. http://www.blogger.com/profile/4269681

58. Andy Miah, "Rethinking Enhancement in Sport," in William Sims Bainbridge and Mihail C. Roco (eds.), *Progress in Convergence* (New York: New York Academy of Sciences, 2006, pp. 301–320).

59. Mihail C. Roco and William Sims Bainbridge (eds.), *Nanotechnology: Societal Implications, Maximizing Benefits for Humanity* (Arlington, VA: National Nanotechnology Coordination Office, 2005, pp. 63–64).

Chapter 5

Cognitive Technology

Since the dawn of history, humans have sought to understand and to command their world. Archimedes is reputed to have said that he could move the world, if he had a lever long enough and a fulcrum strong enough. That lever is converging technologies, and the fulcrum is converging sciences. To move the Earth, Archimedes recognized, one needs a cosmic place to stand. To move the human mind, it was long thought, requires a transcendent standpoint. Today, however, the mysteries of the mind are gradually being unraveled by a multidisciplinary movement called *cognitive science*. The groundwork is being laid for a host of technologies based in its discoveries.

THE TWO FACES OF COGNITIVE SCIENCE

There are at least two very different ways to define cognitive science: as a domain of research and as a social movement with a distinctive ideology. It is as a domain of research that "cog-sci" converges with nanotechnology, biotechnology, and information technology to create new cognitive technologies. Nevertheless, we need to understand the movement as well if we are to have a clear perspective on what cognitive science truly means.

The domain of research covers all systematic studies of human, animal, or machine intelligence, including perception and emotion as well as cognition. The website of the Cognitive Science Society and the front page of its journal, *Cognitive Science*, list seven fields: artificial intelligence, linguistics, psychology, philosophy, neuroscience, anthropology, and education. The last field—education—stresses research on how people learn as much as how cognitive science is taught. In *The Stanford Encyclopedia of Philosophy*, Paul Thagard stresses that cognitive science should be a convergent field, rather than a mere collection of separate disciplines:

> In its weakest form, cognitive science is just the sum of the fields
> mentioned: psychology, artificial intelligence, linguistics, neuro-

science, anthropology, and philosophy. Interdisciplinary work becomes much more interesting when there is theoretical and experimental convergence on conclusions about the nature of mind. For example, psychology and artificial intelligence can be combined through computational models of how people behave in experiments.[1]

As a social movement, cognitive science offered a particular theory of the human mind, rooted in the artificial intelligence work of Marvin Minsky, Allen Newell, and Herbert Simon, plus the generative grammar of Noam Chomsky in linguistics. Many would want to call the theory the "cognitive science theory of mind," but that would cause confusion when we consider the broader definition of cognitive science as a diverse field having many theories. So, I will call it the *structural–computational theory of mind*, emphasizing its main features.

The structural–computational theory postulates that humans have clear and well-defined concepts that can be connected to one another by straightforward, logical derivations, based on manipulating symbols.[2] Many of the concepts define categories that are arranged in a hierarchical manner. For example, we possess a high-level or abstract concept of *animal*. Animals have certain attributes—such as locomotion, skin, and a mouth—that may differ from one kind of animal to another. When asked to imagine an animal, we imagine it having locomotion, but of what kind? The typical mammal walks, but birds fly and snakes slither. The skin of a mammal is covered with hair, and the skin of a bird with feathers. A mammal's mouth has lips, whereas a bird has a beak.

In artificial intelligence, these attributes for which alternatives exist are sometimes called *slots*, indicating that one may place different concepts into them, giving rise to definitions of different things. Literally "mouth" is a slot in the concept *animal* where one may place "lips," giving a crude definition of mammal, or "beak," defining birds. Within the category of birds are many subcategories, many of which are distinguished by having different concepts in a given slot. Here, we have just conceptualized concepts from the top down, as if scanning downward in a pyramid of connected concepts from very general ones to very specific ones.

Alternatively, simple concepts can be connected to make complex ones by a process called *chunking*. I recently went from my home in Arlington, Virginia, to a conference in Lincoln, Vermont, but I did not leap from my home directly into the conference room. Rather, I made a plan of four steps. To go to any far destination, I first take a taxi to National Airport in Arlington. The second leg of this trip was a plane flight from Arlington to Burlington, Vermont. The third leg was a van ride from the Burlington airport to a hotel in Middlebury. The fourth and final leg of the journey was a van ride from the Middlebury hotel to the Lincoln retreat. Thus my plan required chunking together four legs to make one itinerary.

Importantly, the structural–computational theory assumes people follow *logical rules* in mentally manipulating concepts, similar to the rules of formal logic. This theory naturally fits the way many computer scientists and information scientists think. Before the emergence of modern computers, information science largely consisted of librarianship and lexicography. Librarians developed classification schemes, such as the Dewey Decimal System and the Library of Congress System, and lexicographers wrote dictionaries containing formal definitions of all the words. This formal perspective quickly gained dominance in the new field of artificial intelligence, perhaps because it promised a quick way of developing highly intelligent machines. But is this the natural way of thinking for human beings?

I think the answer is "partly," but here is an argument for why the answer might be "no." Many kinds of intelligent behavior do not make use of words at all, so definitions may be irrelevant, or we may use nonverbal definitions that have very different properties. For example, we recognize the faces of friends and family members instantly, without verbally thinking through the list of attributes of the given face, hair color, age, and the like. Likewise, we perform many routine tasks each day without mentally naming their elements or planning how to use them, such as brushing our teeth or tying our shoes. The very fact that it takes several years of schooling to get someone to use a dictionary effectively or to do mathematics suggests that formal definitions and logic are not entirely natural.

The structural–computational theory supported the classical approach to artificial intelligence, which was formalistic, rule-based, and symbolic. Another, very different approach also existed, however—namely, *neural networks*. The human brain consists of literally billions of nerve cells (neurons), each with perhaps a thousand connections to other neurons. In computer science, neural nets are systems of connections between memory registers that are comparable to the connections between neurons, storing information not as symbols but rather as numbers representing the strength of each connection. Learning entails the strengthening or weakening of connections, based on reinforcement or error correction. I have programmed neural networks dozens of times over the past 20 years, and I have found them to be an effective approach for machine learning by trial and error. Yet, to the extent that concepts arise at all in neural networks, they are fuzzy, probabilistic, and overlapping.

The idea of neural networks is really not new, but it has experienced a remarkable roller-coaster of ups and downs over the decades, which suggests it is as much a social movement (and hence competes with other trends) as it is an empirically valid technical discipline.[3] The United States invested in military neural net research at least as early as the mid-1950s, hoping to build electronic devices that could, for example, recognize enemy tanks in the field and direct missiles to them.

In 1969, Marvin Minsky and Seymour Papert published *Perceptrons*, a highly influential book that strongly implied that neural nets were a blind alley.[4] Minsky himself had originally worked with neural nets while he was a student of behaviorist psychologist B. F. Skinner around 1951, but he quickly switched to the opposite intellectual orientation, espousing a symbolic approach to artificial intelligence.[5] A Freudian psychoanalyst might be forgiven for branding this change in attitude an instance of the oedipal complex. Indeed, Skinner was a rather cold fish who did little to boost his students' egos, as I found in my conversations with him. I can remember standing in line with him at the bank, and all I could get him to talk about was how reinforcement learning got people to stand in lines and wait for rewards.

Minsky's defection was highly unfortunate, because even very simple neural nets can do useful work by simulating the conditioning of animals by rewards and punishments. For example, Minsky could have worked with Skinner's friend and colleague, George Homans, who was developing a theory of human social behavior that could readily have been simulated with even the simple neural nets that could have been built in the 1950s.[6] I can remember standing with Skinner and Homans, waiting for the elevator in Harvard's William James Hall, when Skinner actually betrayed a hint of sentiment toward his friend Homans.

"George, I have recommended you for membership in a London club, so you could stay there the next time you visit England," Skinner reported.

Homans replied dryly, "I don't think I'll be going to London again."

Minsky might have had a tough time working for these two emotionally distant and proud men. Homans preferred to express his feelings in formal aphorisms, of which he coined hundreds. He came from a wealthy background, and this fact shaped his experience of the Depression, as he told me: "The Great Depression was great, if you had money." Another time, after sitting through a lecture by a neo-Marxist on the cathartic effect of revolutionary violence, Homans told me, "Violence is fine, if you win." The top ranks of academia are extremely competitive, and Minsky, like Skinner and Homans, gained status by deflating the reputations of people who followed a different approach.

The book by Minsky and Papert asserted that a certain kind of neural net cannot solve a simple problem sometimes called XOR, the "exclusive-or" problem in formal logic. This means learning to respond "1" to two inputs X and Y when X = 1 and Y = 0, and when X = 0 and Y = 1, but to respond "0" when X = Y, either both 0 or both 1. *Perceptrons* does not quite say so, but the reader quickly gets the impression the authors have mathematically proven that all neural nets have this defect and, therefore, that neural nets are of little value in general. According to Michael Swaine, Minsky and Papert were probably aware that a U.S. Army-funded neural net project had already solved the XOR problem.[7] In any case, academic computer science practically abandoned neural net research, apparently in great part as a response to Minsky's immense prestige and a wave of enthusiasm for his very different approach.

In the 1980s, several independent researchers developed solutions to the XOR problem that worked very effectively in simple computer programs. These researchers formed the basis of a new social movement, often called *connectionism*, that sought to replace symbolic approaches by neural nets as the dominant form of artificial intelligence research.[8]

Frankly, neither the structural–computational approach nor the connectionist–neural net approach has been particularly successful in emulating human intelligent behavior. Each can solve certain "toy problems" and has some limited realm of practical applications. One might even argue that each is a correct model of the human mind but simply applies to a different level. At the level of the neurons in the brain (a lower level), the neural net approach is a good metaphor, although neural net computer algorithms probably do not closely resemble how actual nerve cells operate. Presumably, concepts arise in the interactions of hundreds and thousands of neurons, but they may not have clear boundaries or definitions as the structural–computational approach expects. Therefore, when I refer to *cognitive science*, I mean the coming convergence of both of these approaches, along with other approaches including cognitive neuroscience, rather than just a single theory.

COGNITIVE CONVERGENCE

Much cognitive science research has been carried out in the traditional manner in which psychological research is generally done—that is, researchers conduct experiments that present people with various stimuli and situations and then observe the participants' behavior. For example, a century ago, Freud's disciple Carl G. Jung developed the method sometimes called *free association*.[9] Jung had developed a list of 100 words he would say to his research subjects, one at a time. Here are the first 10: head, green, water, sing, dead, long, ship, pay, window, and friendly. The subject was told, "Answer as quickly as possible the first word that occurs to your mind." Jung would not only write down the person's response, but also note how long it took him or her to respond. For example, Jung said "head" to one normal subject, and 1.8 seconds later the person responded "foot." Neurotic subjects often gave distinctive patterns of answers, and sometimes a long delay passed before the answer came. This delay is technically called *latency of response*, and a large latency probably indicates that the person is performing an unusual amount of cognitive work to develop the response.

Psychologists differ in how they interpret this cognitive work in different situations. For example, a psychoanalytically oriented researcher might expect large latencies for words connected with desires the person was trying to repress. If a man associated "mother" with "sex," for example, he might struggle over the response, ultimately either responding "sex" rather late or even saying some other apparently irrelevant word such as "sixty" that concealed his real response, again after a delay while mentally searching for the other word. People with cognitive impairments might have high latencies in general or high latencies for words connected to concepts they had trouble articulating.

In 1966, Saul Sternberg at Bell Telephone Laboratories used the then-new computer technology to explore human short-term memory by measuring latency of response to memory tasks.[10] Sternberg's experiment is so elegant, and gives such interesting results even for just a single research subject, that I have twice replicated it in my own computer programs.[11]

In this type of study, the computer flashes strings of randomly selected digits or letters on the screen, from one to six digits long. My most recent program arranges the strings in many sets, with 120 strings in a set. Each string avoids using a digit or letter twice. The research subject needs to go through all 120 strings of a set during one session. Suppose that one string is "14972." These digits will flash on the screen briefly one at a time, and the person is supposed to remember them. Then, after a pause, a target digit appears inside a box—say, "8." The user is supposed to indicate whether that digit was a member of the string by using the computer's mouse to click "YES" or "NO." In this case, the correct answer is "NO." The computer records how long the response took and whether it was correct, and the machine keeps going until the subject has correctly answered 120 times. In each set of 120 strings, the target digit is a member of the string in 60 cases and not a member in 60 cases. In each group of 60, there are 10 examples where the string is one digit long, 10 where it is two digits long, and so on, up through 10 examples where the string is six digits long.

The first thing to note about this experiment is the remarkable fact that normal human beings are actually able to perform this task! Somehow, we can hold a string of digits in memory, give a response, dump the string out of memory, and get ready to memorize a new string. After a session, the individual probably won't remember any of the strings of digits, so his or her mind won't be cluttered with meaningless numerical trash. This result illustrates the fact that part of human memory remains devoted to the present time— NOW—and a present task, until it has been completed or abandoned. This part of memory is called *short-term memory* or *working memory*.

Some cognitive scientists, when asked "What is consciousness?", might well answer, "Short-term memory." Many people believe that consciousness is a mystery, having something to do with the immortal soul, but one could just as easily argue that it is nothing more than the cognitive part of the human brain's reaction to a mixture of sensory impressions and short-term memory. Anyone who thinks cognitive science has ignored consciousness is mistaken, because much research on this topic can be found under the headings of short-term memory, working memory, and attention.

In 1956, George Miller noted the remarkable fact that humans can hold about only seven items in short-term memory at once.[12] He suggested that the way we can handle more information is first to chunk items together and then to treat this chunk as a new unit. Thus was born the concept of *chunking,* which has proved so important in cognitive science. Seven items seems about the minimum number required for intelligence, and the small capacity of our short-term memory implies we are innately very stupid.

I once met a Harvard student who could carry out long division mentally, handling many digits at once. Clearly, there was something very different about his mental apparatus. It is often thought that the people who have a very distinctive mental talent are otherwise subnormal, *idiot savants*, but this student's verbal behavior and Harvard grades were, in fact, above average. It may be that many above-average people possess hidden mental talents, only some of which have been formally identified, but these individuals either actively conceal those abilities or are not consciously aware of them.

Sternberg's experiment was a first attempt to understand how short-term memory worked, given that he did not know where it was located in the brain or how to inspect it even if it were permissible to open someone's head and look inside. He imagined two different ways that a string of digits might be inspected in short-term memory: one at a time or all at once. Perhaps people have little memory boxes, with room for one digit or one idea in each box. A person memorizes a string by putting one digit in each box: 1 4 9 7 2 . Then comes the target digit, 8. The person looks in each of the five boxes in turn, fails to find "8" in any of them, and reports "NO." This technique requires the person to scan through the entire memorized string. But suppose the target digit had been "4"; then the person would have needed to look in only the first and second boxes and thus could respond "YES" more quickly. This gave Sternberg a way of determining whether people scan through the units of their short-term memory when looking for something or whether they can see the whole short-term memory in a single mental glance. The results of his classic experiment showed that people respond more quickly (with lower latency) when the target is a member of the string, thus support-ing the scanning model of information retrieval from short-term memory.

When I tried the experiment on myself, using the software I pro-grammed following Sternberg's design, I got the same results. If a string of digits was long and the target was in the string, my latency was lower than if it was not in the string. Also, my results illustrated Miller's point, because I made a number of errors for six-digit strings (near the seven-item limit of short-term memory), and even a few for five-digit strings.

Then a colleague of mine confided in me that she possessed an *eidetic memory*, sometimes called *photographic memory*, which has very different short-term capabilities from "normal" people's memories. Her results during the experiment were astonishing. First, she never made any errors. My soft-ware allowed the user to adjust the speed at which the digits were flashed, and she cranked it up much faster than I could handle. Even so, she always gave the correct answer. What's more, her response time latency showed no consis-tent relationship to whether the target digit was in the string or not. Person-ally, I had a small short-term memory and scanned it somewhat erratically when retrieving information. My colleague's short-term memory seemed to be built on different principles from mine, much more capable and reliable, as if we were from different planets.

This is a sterling example of the tantalizing excitement of traditional cognitive science. We gain knowledge about how the human mind works, but only with extreme difficulty. Then, just when we stumble over a great mystery, we run into research barriers that prevent us from understanding it. The greatest of these mysteries for cognitive science is probably the problem of how neural connections give rise to concepts. That is, how do you integrate structural–computational theory with neural net theory, or dictionary defini-tions of concepts with the neurobiology of the brain? Given that computers have proven useful in doing cognitive science research for the past four decades, and that the brain is clearly a biological organ, it would seem reason-able that convergence of the NBIC fields will be hugely advantageous for making progress in cognitive science.

It is important to realize that cognitive science does not exclude the study of emotions, but rather recognizes that thought and feeling are intimately connected. Emotions motivate cognitions, cognitions can trigger emotions, and observations of a person's behavior reveal cognitions and emotions interacting intimately to produce actions and reactions. At the Massachusetts Institute of Technology, Rosalind Picard has long been advancing a research program in *affective computing,* with the aim of giving computers and robots the ability to respond correctly to human feelings. Recently, in collaboration with Rana el Kaliouby, she has been developing a wearable system that can monitor an individual's expression, inferring cognitive and emotional states from them (Figure 5–1). One aim of this research is to develop assistive technologies that can help individual humans who lack a sensitivity to other people's emotions. The broader research involves biotechnology sensors as well as information technology. Of course, emotions are a biological as well as cognitive phenomenon, so the research is highly convergent.[13]

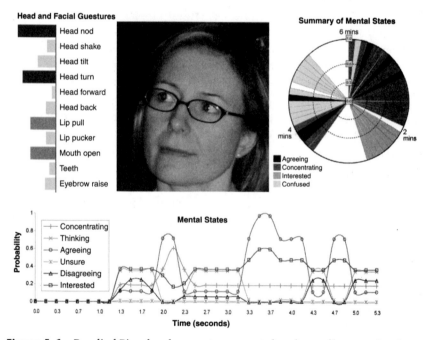

Figure 5–1 Rosalind Picard and computer-generated analyses of her emotional expressions. Picard is a pioneer of "affective computing" research in bio-info-cogno convergence enabling computers and robots to take account of human feelings, and contributing to research on the differences among humans in their social and emotional abilities.

Undoubtedly the most significant new computerized tool for cognitive science research on humans is functional magnetic resonance imaging (fMRI), which can produce motion-picture images of the human brain at different points in time, highlighting which parts of the brain are active during different mental tasks. Michael Gazzaniga, who works at the University of California at Santa Barbara, is among the leaders of the field and the editor of a massive survey of cognitive neuroscience.[14] In 1999, Gazzaniga won a $5 million grant from the National Science Foundation to establish the National Functional Magnetic Resonance Imaging Data Center. This cyberinfrastructure comprises "a shared database in the field of cognitive neuroscience, whose goal is to understand the nature of thought by examining images of brain activity under a range of circumstances and conditions."[15]

Computer entrepreneur Ray Kurzweil has suggested that magnetic resonance imaging (MRI) could be used to read out the neural structure of a person's brain, which then could be simulated inside a computer.[16] Unfortunately, the spatial resolution of MRI is currently far too poor to reveal how individual neurons connect to one another. A voxel (volume pixel) of one cubic millimeter is considered high resolution today, but it can contain as many as 100,000 neurons. We can improve the resolution of MRI by increasing the magnetic field strength, which is typically 1 to 3 teslas. A tesla is 20,000 times the Earth's magnetic field, and the U.S. Food and Drug Administration currently requires special permission for research involving field strengths in excess of 4 teslas. MRI equipment also employs radio waves. The combination of fluctuating magnetic fields and radio waves can cause *peripheral nerve stimulation*, in which the person's muscles start twitching and he or she experiences disturbing sensations. At very high strength, they would cook your brain much in the same way that a microwave oven cooks food.

Although laboratory methods are being developed to trace the actual connections among neurons, this is not a simple business. The root-like structures called dendrites that connect synapses to the neuron cell body can be extremely difficult to disentangle in slices of brain tissue taken during an autopsy, let alone in the living brain. Human neurons are not hard-wired, and it is quite uncertain how much we would need to know about the fine structure of synapses, surrounding glial cells, and neurotransmitter receptors to understand how any group of neurons really interacts. A better course may be to observe how electrical potentials or local metabolisms vary dynamically as a person thinks, perceives, and acts. To this end, some experimenters are working with voltage-sensitive fluorescent dyes, while others are using diluted solutions of calcium indicators.[17] Most recently, researchers at the University of California at San Diego have achieved very-high-resolution, controllable images of neurons using nanoscale fluorescent quantum dots.[18] These methods work only on relatively small exposed surfaces of the brain and are primarily limited to research on animals.

Rodolfo Llinás, chairman of the Department of Physiology and Neuroscience at New York University, thinks he has found a new nanotechnology approach to understanding how neurons interact in the living brain. As he and his colleague Valeri Makarov explained in the first *Converging Technologies* report, it may be possible to feed benign nanoscale wires through the circulatory system into the brain, without obstructing the flow of blood. Many nanoscale wires would be bundled together, but then would fan out in a section of the brain to monitor the electrical behavior from each member of a group of neurons that interact with one another. The animal or person under study would perform actions or receive stimuli, and the output from the system would provide invaluable information about how the brain behaved under those circumstances. When the research session was over, the nanowires could be withdrawn without damage to the brain. This is a radical idea, and many technical hurdles would need to be overcome before it could be realized. Llinás and Makarov say, "We propose that this approach would be very helpful in human capacity augmentation and will yield significant new information regarding normal and abnormal brain function."[19]

Wolfgang Porod, director of the Center for Nano Science and Technology at Notre Dame University, explained to attendees at both the second and third Converging Technologies conferences how his research team combined all four NBIC fields to develop artificial retinas, modeled on biological retinas, that could "see" at a much wider range of wavelengths (colors) than the human eye.[20] The retina does a considerable amount of computing before it sends visual information to the brain—for example, detecting as many as a dozen main features in the image simultaneously. It also seems to compensate for the periodic rapid eye movements called *saccades*, blanking part of its output so that the brain will not be confused by the sudden changes.

One fascinating biotechnology research method used by the Notre Dame team involves a virus that has been genetically engineered to fluoresce green, allowing it to travel from one neuron to the next and thereby tracing out the neural circuit. The retinas qualify as nanotechnology, because some components must be considerably smaller than 100 nanometers to do their job. The biologically inspired circuitry of these visual computers follows a principle called CNN (cellular neural/nonlinear network) design that is relatively new for computer science; it represents a convergence of neural nets with cellular automata. By duplicating the behavior of a biological neurocomputer on a specially manufactured silicon chip, the retinas test scientists' understanding of how human vision works, thus contributing to the advance of cognitive science.

THE PREHISTORY OF COGNITIVE TECHNOLOGIES

Just as some of the accomplishments of chemistry were achieved centuries ago, such as the creation of steel by ancient metallurgists and the development of concrete by Roman builders, some cognitive technologies date from bygone times. One could argue that *counting*—a practice we take for granted—was one of the first and most effective cognitive technologies; it was later followed by simple arithmetic, and then by the sophisticated geometry of the ancient Egyptians, Babylonians, and Greeks. Writing could certainly be described as an ancient cognitive technology, although it appears to have been based on precious little linguistic science. Despite those developments of the ancient world we would like to count as pioneering cognitive technologies, previous civilizations certainly lacked real cognitive science, and thus they often went astray in the search for ways to improve the human mind.

Cognitive technologies seek to augment the capabilities of the human mind, so their success inevitably requires a proper understanding of how the mind works.[21] For most of human history, this understanding was lacking; indeed, people possessed a whole menagerie of illusions about themselves. Too much investment of hope and resources in cognitive technology, before any true understanding of the mind has been achieved, is disastrous. It not only wastes effort but also inhibits progress in other areas of science and engineering. Long before science could frame successful theories of reality, for example, religion convinced people that it did so, both because faith offered hope and because supernatural claims are not easily refuted empirically.[22]

Ancient Egypt is famous for its early technological development, represented not only by the great pyramids but also by the fruitfulness of its agriculture, advances in medicine, and the independent invention of writing. Many reasons may explain why Egyptian progress essentially stalled soon after the pyramids were built, but one factor may have been the development of religious technologies of immortality, including cognitive technologies.

No one today believes that mummification and the associated rituals actually grant the individual immortality. But for the ancient Egyptians, elaborate preparation of the dead body transformed it into a proper vehicle to transport the soul to the afterlife.[23] Indeed, the Egyptians had a very complex theory that the human personality was a collection of separable parts that needed to be reassembled after death. The ritual called "opening the mouth" was intended to insert part of the spirit back into the body. By one estimate, the sands of Egypt hold 100 million mummified bodies, and vast wealth was poured onto those sands in a vain attempt to live forever. In consequence, both less material wealth and less human talent were available for more realistic fields of science and engineering.

A comparable problem has plagued Asian societies. Variants of Hinduism, Buddhism and other eastern religions practice what might be called *spiritual magic*. Through rigorous meditation, yoga exercises, and numerous rituals, people seek to gain enlightenment and transcendent powers. These *technological religions* believe they can transform the human sprit through carefully controlled effort. Again, the result is to drain talent and investment away from more realistic technologies, and to impose upon the society ideologies that inhibit real progress. Not all Asian faiths directly encourage magical thinking, of course; Confucianism, which was chiefly an elite ethical code, did not. However, most people in Asian societies participate in multiple religio-magical traditions simultaneously. By contrast, in the Judeo–Christian–Islamic tradition, one or another faith has often achieved a near-monopoly over others in a particular society.

For a century, sociologists such as Max Weber have argued that a decisive historical characteristic of Christianity was its suppression of magic.[24] Of course, Christianity has appeared in diverse forms and has known many heresies over the centuries. Nevertheless, its mainstream version has relied on the divine being, Jesus Christ, to save souls, rather than on methods practiced by mortals. People can lead righteous lives, and they can pray for God's help. But Christianity does not encourage believers to learn techniques designed to improve their souls, minds, or personalities. Especially after Protestantism downgraded monasticism and the priesthood, Christianity has left proponents of secular science and engineering free to explore and master the real world, which they have done with great and continuing success.

In the twentieth century, the groundwork for realistic cognitive technologies was laid by the establishment of psychology as an academic discipline, the emergence of neurobiology, and the first primitive steps in the development of artificial intelligence. Psychoanalysis and the swarm of psychotherapies that arose in its wake were probably a false start in the founding of real cognitive technologies, but their emergence signaled the pent-up demand for cognitive technologies that exists in modern society. Unfortunately, they were not based on either a solid tradition of prior scientific research or careful studies of their own effectiveness.

Today we stand at the threshold of a true understanding of how the human mind works. Optimist that I am, I think we may have a full understanding by the end of the twenty-first century. However, even if it takes us another two centuries rather than one, we already understand much about the mind. For the first time in human history, effective cognitive technologies based on solid scientific research have begun to appear. The first applications may be modest, and they certainly will not confer immediate immortality or freedom from fear, confusion, and sin. But they will enhance our lives and feed back into the process of scientific–technical development to achieve still more progress.

Unfortunately, this transition will be painful for many people, as we have to deal with the remnants of our many illusions inherited from previous centuries and millennia. The western religious traditions that protected us from the folly of supernatural cognitive technologies did so by means of counteractive superstitions. Some scholars argue that science and technological development are rooted in western religious traditions, yet none can deny that science and religion have existed in a tense relationship.[25] For two reasons, that relationship may erupt into overt conflict during this century.

First, the heart of Christianity is a special conception of the nature of a human being. Much is made of the Christian conception of God, a loving but demanding creator mysteriously manifesting as a trinity, who has the power to intervene in individual lives but gives humans considerable freedom. Less is said about the Christian theory of the human mind, but it may be even more important. The doctrine of the immortal soul is notable not only for the concept of immortality, but also for its conceptualization of a person as a spiritual and moral unity that transcends the material world. The emergence of psychiatry in the nineteenth century raised the issue of how to reconcile this conception with the demonstrable fact that injury to the brain could cause radical changes in personality and behavior.[26] In 2006, columnist George Will confronted this painful fact in an essay he wrote about his mother's dementia, when she died at the age of 98:

> It is said that God gave us memory so we could have roses in winter. Dementia is an ever-deepening advance of wintry whiteness, a protracted paring away of personality. It inflicts on victims the terror of attenuated personhood, challenging philosophic and theological attempts to make death a clean, intelligible, and bearable demarcation. Is death the soul taking flight after the body has failed? That sequence—the physical extinguished, the spiritual not—serves our notion of human dignity. However, mental disintegration mocks that comforting schema by taking the spirit first.[27]

The modern conception of the brain as a distributed neural network organized in complexly interconnected modules, in which thoughts and memories are lodged in physical structures, could hardly be more different from the notion of a transcendent soul.[28] The gradual but constant progress of artificial intelligence is likely to challenge the traditional religious viewpoint ever more decisively in the coming years. On the op-ed page of *The New York Times*, Yale psychologist Paul Bloom wrote, "The great conflict between science and religion in the last century was over evolutionary biology. In this century, it will be over psychology, and the stakes are nothing less than our souls."[29]

The second reason cognitive science will conflict with faith is precisely the result of convergence. The implicit centuries-long truce that has existed between science and religion was based not only on science's willingness to stay out of religion's home territory, but also on the high degree of specialization in science. An individual scientist could be religious, despite his knowledge of facts in his own area that contradicted traditional religious beliefs, because he could ignore the secularizing influence of the other separate branches of science about which he knew little. By bringing the sciences and technologies together, convergence will leave little room for faith.

In his influential book *Consilience*, Edward O. Wilson wrote about the rapid unification of scientific knowledge occurring today, and wondered whether the natural sciences would be able to unite with the humanities and religion, which traditionally have claimed to understand humanity itself.[30] By presenting a comprehensive model of reality, science will leave religion little scope to exist. The result could be an estrangement between religious and secular groups in society. At the same time, technological convergence will mean that everyone uses the benefits of science. As a consequence, everyone will have more reason to believe in science than in the ancient myths left over from a primitive age in which kings ruled society and, therefore, people imagined the universe must also have a king.

A leading cognitive theory of religion, developed by a number of researchers including Rodney Stark, Roger Finke, and myself, derives faith in the supernatural from the fact that the natural world does not provide humans with all the rewards they desire.[31] In the absence of a highly desired reward, such as eternal life, humans will accept beliefs that "posit attainment of the reward in the distant future or in some other non-verifiable context . . . Compensators are postulations of reward according to explanations that are not readily susceptible to unambiguous evaluation. . . . [Religions are] systems of general compensators based on supernatural assumptions."[32] Before most people will be willing to forsake religious faith, science and technology will need to compensate them for their psychological loss, both by providing an array of exceedingly valuable new technological rewards and by offering a personal science that cherishes and celebrates the uniqueness of the human individual.

Clearly, cognitive technologies are already entering our lives. The revolution has begun with humble tools such as the spell checker in our computer's word processor and the search engines we use to navigate the World Wide Web. As the technology changes, so humans will also change, just as we have done so many times in the past when our own creativity has transformed the nature of our lives. Ideally cognitive technologies will improve human life sufficiently that wishful thinking will no longer be necessary. Ancient religions served humanity long and sometimes well, but now we must abandon them as we follow the biblical directive of John 8:32: "And ye shall know the truth, and the truth shall make you free."

NEUROTECHNOLOGY

Over the coming century, biotechnology may have a huge impact on human cognition. From the dawn of history, humans have used naturally occurring biochemicals to alter brain function: nicotine in tobacco to calm the nerves, alcohol in wine to release inhibitions, and caffeine in coffee to energize action. Arguably, one of the most harmful technological innovations in human history was the development of industrial distillation of alcohol in the nineteenth century, leading to the cheap production of "hard liquor" and the mass production of alcoholics.[33] Whatever songs one may sing about the glories of rum, uncounted millions of people have suffered from alcoholism, through either their own addiction to alcohol or that of a key family member.

Modern societies tend to control substances that experts believe are harmful or that political leaders believe undermine the traditional culture. For this reason, it is likely that new mind-affecting drugs will be controversial and possibly banned. The classic case is the U.S. experiment with prohibition of alcohol in 1919–1933, which imposed on the entire nation the temperance views of a largely Protestant social movement.[34] Although this experiment is widely considered a failure, in fact a number of alcohol-related social problems declined during Prohibition—most obviously, deaths from cirrhosis of the liver. The death rate from this cause was 15.1 persons per 100,000 population in 1916, shortly before Prohibition began, but only 8.8 deaths per 100,000 population in 1923, after it had become well established. Over the entire duration of Prohibition, more than 88 lives were likely saved for every 100,000 urban dwellers owing simply to the prevention of cirrhosis of the liver.[35]

A second classic case is the rise and fall of the Psychedelic Movement in the 1960s and early 1970s. Although I was not personally involved in this movement, I observed it from a rather good vantage point. I served as teaching assistant to Dr. Paul A. Walters, chief psychiatrist in Harvard University's health service, who taught a popular course each year on what he believed were the virtues of psychedelic drugs. Paul was a mixture of the conventional and the radical. His psychiatric theories were compatible with radical redefinitions of mental health and the use of intense intimacy to achieve it. His course lectures argued that each person needs a set of defining experiences to establish an individual identity, and drew on the psychoanalytic theories of D. W. Winnicott that through play infants discover the boundaries between the self and the parent.[36] The readings assigned for the class included three books by Carlos Castaneda about his alleged but probably fictional relations with a Yaqui sorcerer who gave Castaneda psychoactive drugs to transform his perception of reality.[37]

Harvard was the intellectual center for the Psychedelic Movement of the 1960s, largely because of the influence of personality and social psychologist Timothy Leary, and his associate, Richard Alpert. In 1957, before getting involved with drugs, Leary published an excellent scientific monograph presenting his own systematic approach to personality research, and clearly he had a fine mind.[38] Three years later, while traveling in Mexico, he experimented with "magic mushrooms," which were considered sacred in local pre-Christian culture; these mushrooms contained a mind-altering substance, psilocybin. In 1966, Leary founded the League for Spiritual Discovery (LSD) on the premise that a comparable chemical, lysergic acid diethylamide (LSD), could provide spiritual insights from genuine transcendental experiences. Needless to say, this claim was highly controversial and drew the ire of authorities. The Beatles have said that their song "Lucy in the Sky with Diamonds" does not refer to LSD, although every fan thinks it does, and rationalist opponents of Leary suggested that LSD stood for "Let Sanity Die."

Leary was fired from Harvard in 1963. The only time I met him was in 1983, when a graduate student of mine organized a spectacular event in Harvard's Sanders Theater to celebrate the twentieth anniversary of that historic event. Sanders Theater is a famous architectural curiosity, modeled on the Sheldonian Theatre at Oxford University. The Sheldonian was built in 1668, designed by Christopher Wren, and placed the stage nearly in the center of the audience. Sanders Theater is housed in Memorial Hall, which was completed exactly two centuries after Wren's structure and sought to commemorate Harvard men who had died in the Civil War. During the 1983 event, Leary and Alpert both paraded around the stage of the theater, to the applause of a thousand admirers who clearly thought they were heroes rather than villains.

It is easy to dismiss Leary's later career as crazy or criminal, and it is true that he served time in prison for using and distributing illegal drugs.[39]

However, a number of his contemporaries—notably, Thomas Szasz, Thomas Scheff, R. D. Laing, Erich Goode, and Andrew Weil—argued that "reality" was partly a matter of social definitions and thus a political issue during this period.[40] In 1971, twin brothers Zhores and Roy Medvedev published *A Question of Madness*, describing how the Soviet Union locked Zhores in a mental hospital to discredit his criticism of Soviet science policy.[41] Some might argue in a similar manner that Timothy Leary had been imprisoned by the United States to discredit him. Over the decades, in fact, anthropologists had come to a consensus that mental illness was at least partly an arbitrary social definition, and that use of psychedelic drugs could be legitimate within a religious or spiritual cultural context.[42]

Leary's famous slogan, "Turn on, tune in, drop out," recognized that the Psychedelic Movement was in great measure a radical revolt against the demands for disciplined labor that capitalism placed on workers. Thus the question remains open as to whether his imprisonment was simply an appropriate law enforcement judgment, a wise act to protect the public from drugs that caused madness, a tyrannical act intended to defend the power of the capitalist ruling class, or a philosophical mistake based on an ignorant interpretation of the psychedelic ideology. This multidimensional quandary is still relevant today, because a new generation of advocates is now arguing that very different categories of chemical substances should be freely available to expand the powers of the human mind. It will be interesting—to say the least—to see whether governments seek to suppress the use of these substances or whether they come to some accommodation with the proponents of their use.

Zack Lynch (Figure 5–2a), director of an information service called Neuro-Insights, participated in both the second and third Converging Technologies conferences. He argued that *neurotechnology* will be the dominant wave of rapid technological advance over the next 50 years, comparable to the information technology revolution of the past half-century, but that this transformation will not be possible without NBIC convergence. In his vision, we will gain new tools for our brains that will enhance our ability to accomplish goals. Thus much of neurotechnology will harmonize with societal values, rather than challenging them as Leary's work did. In addition, Lynch categorizes neurotechnologies in terms of their technical approach toward problems that are predominately medical in nature:

- Neuropharmaceuticals: pharmaceuticals and biopharmaceuticals targeting the nervous system
- Neurodevices: medical devices, electronics, and software for nervous system disorders

- Neurodiagnostics: brain imaging, molecular diagnostics, and informatics systems[43]

Lynch further classifies neuropharmaceuticals (or *neuroceuticals*) into four groups:

- *Cogniceuticals* will help failing memory, mild cognitive or attention impairments, and insomnia.
- *Sensoceuticals* will treat deafness, blindness, epilepsy, tremor, and chronic pain.
- *Emoticeuticals* will mitigate depression, mania, and anxiety.
- Complex combinations of neurotechnologies will be devised for psychosis, addiction, and stroke.

Lynch observes that neurotechnology could become highly controversial when it seeks to enhance the performance of normal individuals, rather than treating disease and disability. Like many other writers who have addressed such issues recently, he cites the influential 2003 report of the U.S. President's Council on Bioethics, *Beyond Therapy*, which explores the ethics of human enhancement. The report defines "enhancement" as follows:

> [T]hose well-meaning and strictly voluntary uses of biomedical technology through which the user is seeking some improvement or augmentation of his or her own capacities, or, from similar benevolent motives, of those of his or her children.[44]

Figure 5–2 Zack Lynch (a) and Wrye Sententia (b). These science policy analysts have been examining how cognitive enhancement and neurotechnology could offer vast new opportunities for improving human welfare, while raising questions about protection of individual liberty in an era in which human nature is undergoing a transformation.

The same report also contrasts therapy with enhancement. Therapy is "the use of biotechnical power to treat individuals with known diseases, disabilities, or impairments, in an attempt to restore them to a normal state of health and fitness," whereas enhancement is "the directed use of biotechnical power to alter, by direct intervention, not disease processes but the 'normal' workings of the human body and psyche, to augment or improve their native capacities and performances."[45] Distinguishing the two is often difficult, however, in part because people tend to disagree about the definition of "normal."

Consider two runners who are competing with each other. One has allergies that can be managed by taking antihistamines. The other has weak muscles that could be strengthened by taking steroids. At the present time, many people would say allergies are a medical problem but weak muscles are normal. Based on this perspective, they might conclude that it was ethical for the first runner to take antihistamines, but not acceptable for the second runner to take steroids. As the *Beyond Therapy* report notes, there is something illogical about using a technology to improve the intelligence of a person with a low IQ but not the intelligence of a person with a higher IQ, given that both IQs are just points along a normal distribution.

Wrye Sententia (Figure 5–2b), of the Center for Cognitive Liberty and Ethics, appeared at the same convergence conferences as Lynch. She argued that the concept of cognitive liberty was the proper starting point from which to negotiate between people with different demands:

> Cognitive liberty is every person's fundamental right to think independently, to use the full spectrum of his or her mind, and to have autonomy over his or her own brain chemistry. Cognitive liberty concerns the ethics and legality of safeguarding one's own thought processes, and by necessity, one's electrochemical brain states. The individual, not corporate or government interests, should have sole jurisdiction over the control and/or modification of his or her brain states and mental processes.[46]

This principle potentially contradicts the idea that governments can ban selected mind-altering substances or give a monopoly over their use to licensed professionals such as psychiatrists. In particular, it would have defended Timothy Leary's right to use and share psychedelic drugs, especially as they were not proven to be addictive. To say that someone is addicted to a drug implies that the person has lost his or her autonomous power of will, and thus that the drug has disabled the person's cognitive liberty. Claiming that a person is addicted to a drug, therefore, might give government authorities the right to take a person into custody and then place him or her into prison or into a treatment program. The very concept of addiction has been questioned, however. For example, William McAuliffe of Harvard Medical School and Robert Gordon of Johns Hopkins University have argued that so-called drug addicts want to take the drugs because they cause pleasure, rather than because failure to take them would cause pain.[47] This book is not the appropriate place to decide issues related to the nature of addiction; rather, the point being made here is that it is very hard to justify prohibiting free use of any drug that is not believed to be addictive.

One could argue that Timothy Leary was wrong to use his position of scientific authority to advocate the use of psychedelic drugs. Our culture values freedom of expression, however, and autonomous adults must be capable of deciding for themselves what to believe or disbelieve. Indeed, science itself progresses through debate. Furthermore, many of the present and future neuroceuticals may produce results that are widely regarded as beneficial, and controversies of very different kinds may center on different brain-altering chemicals.

Sententia raises the fascinating issue of memory drugs, chemicals that either help people remember or help them forget.[48] She cites the example of memantine, the first medication developed to combat memory loss in patients with Alzheimer's disease. This drug apparently resists the degeneration of memory, though it may not enhance memory in normal people. Sententia also reports that some medications, such as propranolol, may dull unwanted memories, such as those implicated in post-traumatic stress syndrome. One would think that evolution had already given humans a near-optimal balance between remembering and forgetting, because both are necessary for creative intelligence. Perhaps most humans will decide to leave the memory in their brains the way it is and instead seek to enhance memory primarily through personal, portable information technologies.

THE COMMUNICATOR

A task force of the original Converging Technologies conference, while examining how NBIC might enhance group and societal outcomes, suggested that the greatest near-term development coming out of a union of cognitive science with other fields would be an information technology system to support human interaction they called *The Communicator*. The leading visionary in this group was Philip Rubin (Figure 5–3), who was then on loan from Yale University and was serving as director of the Division of Behavioral and Cognitive Sciences at the National Science Foundation. Phil has since returned to

Figure 5–3 Philip Rubin. As director of the Division of Behavioral and Cognitive Sciences at the National Science Foundation, Rubin made crucial contributions to the original Converging Technologies conference, thereby launching the NBIC effort. One result was his vision of The Communicator, an NBIC system that would allow groups of people anywhere to communicate and collaborate successfully, even across cultural divides.

his regular job as chief executive officer of Haskins Laboratories, a research center devoted to language and its biological basis. Among Phil's claims to fame are his early research on producing recognizable speech with pure sine waves and his ownership of one of the world's largest collections of toy robots.

More than a dozen other top scientists brainstormed with Rubin, with the goal of identifying humanity's greatest need and determining how technological convergence might satisfy it over the next two decades. They judged that good communication, based on solid knowledge, was the key to solving all other important human problems. They saw real potential for nano-enabled information technologies, incorporating biosensor data and designed on the basis of advanced scientific understanding of human cognition, to make The Communicator possible. This development would be an outgrowth of today's cutting-edge wearable or pocket computers with Internet connectivity, and would give the user access to all the world's knowledge, wherever he or she might be at the moment. But The Communicator would be far more than just a pocket electronic encyclopedia:

> The Communicator will also be a facilitator for group communication, an educator or trainer, and/or a translator, with the ability to tailor its personal appearance, presentation style, and activities

to group and individual needs. It will be able to operate in a variety of modes, including instructor-to-group and peer-to-peer interaction, with adaptive avatars that are able to change their affective behavior to fit not only individuals and groups, but also varying situations. It will operate in multiple modalities, such as sight and sound, statistics and text, real and virtual circumstances, which can be selected and combined as needed in different ways by different participants. Improving group interactions by brain-to-brain and brain–machine–brain interfaces will also be explored.[49]

Reginald G. Golledge, a blind geographer, was among the most active contributors to the first two Converging Technologies conferences, and he energetically proposed that an early version of The Communicator should help blind people find their way through the world. The very concept of a blind geographer is remarkable, and Reg made a powerful impression when he spoke at the first conference. He began by placing a graphic on a plastic sheet on the overhead projector, with a piece of opaque paper on top of it. Naturally, nothing projected on the screen. At first, we thought he was one of those coy lecturers who likes to control the audience's attention by gradually withdrawing the paper to reveal key points, one at a time. Then it occurred to us that, being blind, he might not even know the paper was there, blocking the projection. People in the room began fidgeting, until finally one had the temerity to interrupt Reg and inform him that no one could see the graphic. Of course, Reg was fully aware of the fact, and was using it as a dramatic demonstration of how lectures looked to blind people.

Some of Golledge's own research has shown not only that blind people have trouble navigating the environment because they cannot see, but also that they have more trouble learning a route even when they can navigate it.[50] Also, as they walk through a city, even if they find their way, visually impaired individuals will not see signs and other cues that would tell them about potentially useful resources in the neighborhood. If the aim of converging technologies is to enhance human performance, then the abilities of blind people should be an early priority for enhancement. For example, a personal guidance system based in an Internet-connected wearable computer using the global positioning system (GPS) could exchange data with smart signs, locator beacons, home pages of businesses on the street, and "talking neon lights" in an airport terminal that modulate their light with spoken directions to likely destinations that could be heard through the device. Golledge is a committed advocate for convergence, arguing that anyone would benefit from access to a wider range of information resources and perspectives.[51]

Golledge's colleague, Jack Loomis, believes that perception can often be improved by substituting one human sense for another, either when a particular sense has been damaged and information must be presented as sound or touch, or when the information input is so great that multiple senses must be used simultaneously to accommodate it.[52] Two participants in the first NBIC meeting, and one in the second Societal Implications of Nanotechnology meeting, imagined light headgear that could unobtrusively provide information to the wearer, possibly under brain-wave control.[53] For Sherry Turkle, director of the MIT Initiative on Technology and Self, robots and wearable computers will be not merely tools, but rather companions that are sufficiently sophisticated *sociable technologies*.[54] Of course, only if our intelligent machines understand humans very well can they help us communicate on the deepest levels.

CONCLUSION

Earlier in the brief history of the converging technologies movement, the members of the nano-bio-info triad were referred to as technologies, whereas the cognitive area was described as a science.[55] In reality, all four areas include both sciences and technologies—nanoscience as well as nanotechnology, for example—and convergence will require both fundamental research and applied engineering. Cognitive science is a relatively new field, and perhaps only now is it accurate to say that well-grounded cognitive technologies finally exist.

Cognitive science is itself a convergent field, bringing together psychology, neurobiology, linguistics, artificial intelligence, and some aspects of economics and anthropology; it has yet to integrate sociology or political science, but could do so in the future. Integration of cognitive principles into information systems is already taking place (e.g., web browsers, recommender systems, reputation systems, human–computer interaction studies), but research on nano-enabled mobile information systems will open an entirely new realm for such advances. Nano-bio methods promise to unlock the secrets of the human brain in the next very few years, providing models for much more effective artificial intelligences and helping us design all kinds of technology for more effective human use.

Cognitive technologies are science-based methods for augmenting or supplementing human knowledge, thought, and creativity. In the coming decades, they will benefit greatly from convergence with the three other domains of knowledge and capability. From biology and biotechnology, they will gain a constantly improving understanding of the human brain, psychiatric and normal-enhancing medications, and a systematic appreciation of the role of emotions in guiding and energizing intelligence. From information science and information technology will come databases (e.g., bioinformatics, nanoinformatics), new tools for communication, and artificial intelligence to gradually supplement (but never supplant) the power of our own minds. From nanoscience and nanotechnology will come the methods needed for brain research, sensors for capturing new kinds of information about the environment, and the nanoscale components required for truly mobile information processing. Cognitive technologies, in return, will offer the other domains new ways to conceptualize and communicate about their realms of reality.

REFERENCES

1. Paul Thagard, "Cognitive Science," in Edward N. Zalta (ed.), *The Stanford Encyclopedia of Philosophy* (winter 2004 edition), online at http://plato.stanford.edu/archives/win2004/entries/cognitive-science/

2. Allen Newell, *Unified Theories of Cognition* (Cambridge, MA: Harvard University Press, 1990); Herbert A. Simon, *The Sciences of the Artificial* (Cambridge, MA: MIT Press, 1996).

3. John von Neumann, "The General and Logical Theory of Automata," in James R. Newman (ed.), *The World of Mathematics* (New York: Simon and Schuster, 1956, pp. 2070–2098).

4. Marvin Minsky and Seymour Papert, *Perceptrons* (Cambridge, MA: MIT Press, 1969).

5. Daniel Crevier, *AI: The Tumultuous History of the Search for Artificial Intelligence* (New York: Basic Books, 1993).

6. William Sims Bainbridge, "Grand Hopes: Review of Crevier," *Science*, 261:1186, 1993.

7. Michael Swaine, "Two Early Neural Net Implementations," *Dr. Dobb's Journal*, 14(11):124–131, 1989.

8. David E. Rumelhart and James L. McClelland, *Parallel Distributed Processing* (Cambridge, MA: MIT Press, 1986); Philip D. Wasserman, *Neural Computing: Theory and Practice* (New York: Van Nostrand Reinhold, 1989); Philip D. Wasserman, *Advanced Methods in Neural Computing* (New York: Van Nostrand, 1993); Geoffrey E. Hinton, "How Neural Networks Learn from Experience," *Scientific American*, 267(3):145–151, 1992.

9. Carl G. Jung, "The Association Method," *American Journal of Psychology,* 31:219–269, 1910.

10. Saul Sternberg, "High-Speed Scanning in Human Memory," *Science,* 153:652–654, 1966.

11. The first time was reported in William Sims Bainbridge, *Experiments in Psychology* (Belmont, CA: Wadsworth, 1986); the second time is the research reported here.

12. George A. Miller, "The Magical Number Seven, Plus or Minus Two: Some Limits on Our Capacity for Processing Information," *Psychological Review,* 63:81–97, 1956.

13. Rana el Kaliouby, Rosalind Picard, and Simon Baron-Cohen, "Affective Computing and Autism," in William Sims Bainbridge and Mihail C. Roco (eds.), *Progress in Convergence* (New York: New York Academy of Sciences, 2006, pp. 228–248).

14. Michael S. Gazzaniga (ed.), *Cognitive Neuroscience: A Reader* (Malden, MA: Blackwell, 2000).

15. Michael S. Gazzaniga and Daniel N. Rockinore, "Data Are Most Useful When Openly Shared," *Chronicle of Higher Education,* March 16, 2001, p. B13; http://www.cs.dartmouth.edu/~rockmore/Data_Chron.html; http://www.nsf.gov/awardsearch/showAward.do?AwardNumber=9978116

16. Ray Kurzweil, *The Age of Spiritual Machines: When Computers Exceed Human Intelligence* (New York: Viking, 1999, pp. 6, 122–123).

17. Zita A. Peterlin, James Kozloski, Bu-Qing Mao, Areti Tsiola, and Rafael Yuste, "Optical Probing of Neuronal Circuits with Calcium Indicators," *Proceedings of the National Academy of Sciences,* 3619–3624, 2000; Doron Shoham, Daniel E. Glaser, Amos Arieli, Tal Kenet, Chaipi Wijnbergen, Yuval Toledo, Rina Hildesheim, and Amiram Grinvald, "Imaging Cortical Dynamics and High Spatial and Temporal Resolution with Novel Blue Voltage-Sensitive Dyes," *Neuron,* 24:791–801, 1999.

18. Smita Pathak, Elizabeth Cao, Marie C. Davidson, Sungho Jin, and Gabriel A. Silva, "Quantum Dot Applications to Neuroscience: New Tools for Probing Neurons and Glia," *Journal of Neuroscience,* 26(7):1893–1895, 2006.

19. Rodolfo R. Linás and Valeri A. Makarov, "Brain-Machine Interface via a Neurovascular Approach." in Mihail C. Roco and William Sims Bainbridge (eds.), *Converging Technologies for Improving Human Performance* (Dordrecht, Netherlands: Kluwer, 2003, pp. 244–251).

20. Wolfgang Porod, Frank Werblin, Leon O. Chua, Tamás Roska, Angel Rodriguez-Vásquez, Botond Roska, Patrick Fay, Gary H. Bernstein, Yih-Fang Huang, and Árpád I Csurgay, "Bio-Inspired Nano-Sensor-Enhanced NCC Visual Computer," in Mihail C. Roco and Carlo D. Montemagno (eds.), *The Coevolution of Human Potential and Converging Technologies* (New York: New York Academy of Sciences, 2004, pp. 92–109); Gary H. Bernstein, Leon O. Chua, Árpád I. Csurgay, Patrick Fay, Yih-Fang Huang, Wolfgang Porod, Ángel Rodriguez-Vásquez, Botond Roska, Tamás Roska, Frank Werblin, and Ákos Zarándy, "Biologically-Inspired Cellular Machine Architectures," in William Sims Bainbridge and Mihail C. Roco (eds.), *Managing Nano-Bio-Info-Cogno Innovations: Converging Technologies in Society* (Berlin: Springer, 2006, pp. 133–152).

21. Steven Pinker, *How the Mind Works* (New York: Norton, 1997).

22. Rodney Stark and William Sims Bainbridge, *A Theory of Religion* (New York: Lang, 1987).

23. E. A. Wallis Budge, *The Mummy: Chapters on Egyptian Funereal Archaeology* (New York: Biblo and Tannen, 1964); John H. Taylor, *Unwrapping a Mummy: The Life, Death and Embalming of Horemkenesi* (Austin, TX: University of Texas Press, 1966); Ange Pierre Leca, *The Egyptian Way of Death* (New York: Doubleday, 1981); Rosalie David and Eddie Tapp (eds.), *Evidence Embalmed: Modern Medicine and the Mummies of Ancient Egypt* (Manchester, UK: Manchester University Press, 1984).

24. Max Weber, *The Protestant Ethic and the Spirit of Capitalism* (New York: Scribner's, 1958).

25. Richard S. Westfall, *Science and Religion in Seventeenth-Century England* (New Haven, CT: Yale University Press, 1958); Robert K. Merton, *Science, Technology and Society in Seventeenth-Century England* (New York: Harper and Row, 1973).

26. Isaac Ray, *Mental Hygiene* (Boston: Ticknor and Fields, 1863); *Treatise on the Medical Jurisprudence of Insanity* (Boston: Little, Brown, 1871).

27. George Will, "A Mother's Love, Clarified," *Washington Post*, July 13, 2006, p. A23.

28. Steven Pinker, *How the Mind Works* (New York: Norton, 1997); Dan J. Stein and Jacques Ludik (eds.), *Neural Networks and Psychopathology* (Cambridge, UK: Cambridge University Press, 1998); Philip T. Quinlan (ed.), *Connectionist Models of Development: Developmental Processes in Real and Artificial Neural Networks* (New York: Psychology Press, 2003); Thomas R. Shultz, *Computational Developmental Psychology* (Cambridge, MA: MIT Press, 2003); Paul Bloom, *Descartes' Baby: How the Science of*

Child Development Explains What Makes Us Human (New York: Basic Books, 2004).

29. Paul Bloom, "The Duel between Body and Soul," *New York Times*, September 10, 2004, p. A27.

30. Edward O. Wilson, *Consilience: The Unity of Knowledge* (New York: Knopf, 1998); cf. Daniel C. Dennett, *Darwin's Dangerous Idea* (New York: Simon and Schuster, 1995).

31. Rodney Stark and William Sims Bainbridge, *The Future of Religion* (Berkeley, CA: University of California Press, 1985); Rodney Stark and William Sims Bainbridge, *A Theory of Religion* (New York: Lang, 1987); Rodney Stark and Roger Finke, *Acts of Faith* (Berkeley, CA: University of California Press, 2000); William Sims Bainbridge, "Sacred Algorithms: Exchange Theory of Religious Claims," in David Bromley and Larry Greil (eds.), *Defining Religion* (Amsterdam: JAI Elsevier, 2003, pp. 21–37); William Sims Bainbridge, "A Prophet's Reward: Dynamics of Religious Exchange," in Ted G. Jelen (ed.), *Sacred Markets, Sacred Canopies* (Lanham, MD: Rowman and Littlefield, 2002, pp. 63–89); William Sims Bainbridge, *God from the Machine: Artificial Intelligence Models of Religious Cognition* (Lanham, MD: AltaMira Press, 2006).

32. Rodney Stark and William Sims Bainbridge, *A Theory of Religion* (New York: Lang, 1987, pp. 35–39).

33. James S. Roberts, *Drink, Temperance and the Working Class in Nineteenth Century Germany* (Boston: Allen and Unwin, 1984).

34. Joseph R. Gusfield, *Symbolic Crusade* (Urbana, IL: University of Illinois Press, 1963).

35. Rodney Stark and William Sims Bainbridge, *Religion, Deviance, and Social Control* (New York, Routledge, 1996, pp. 84–92).

36. D. W. Winnicott, *Playing and Reality* (New York, Basic Books: 1971).

37. Carlos Castaneda, *The Teachings of Don Juan: A Yaqui Way of Knowledge* (Berkeley, CA: University of California Press, 1968); *A Separate Reality: Further Conversations with Don Juan* (New York: Simon and Schuster, 1971); *Journey to Ixtlan: The Lessons of Don Juan* (New York: Simon and Schuster, 1972).

38. Timothy Leary, *Interpersonal Diagnosis of Personality: A Functional Theory and Methodology for Personality Evaluation* (New York: Ronald Press, 1957).

39. Timothy Leary, *Flashbacks: An Autobiography* (Los Angeles: J. P. Tarcher, 1983).

40. Thomas Szasz, *The Myth of Mental Illness* (New York: Delta, 1961); Thomas Scheff, *Being Mentally Ill* (Chicago: Aldine, 1966); R. D. Laing, *The Politics of Experience* (New York: Ballantine, 1967); Erich Goode, "Marijuana and the Politics of Reality," *Journal of Health and Social Behavior*, 10:83–94, 1969; Andrew Weil, *The Natural Mind: A New Way of Looking at Drugs and the Higher Consciousness* (Boston: Houghton Mifflin, 1972).

41. Zhores A. Medvedev, *The Rise and Fall of T. D. Lysenko* (New York: Columbia University Press, 1969); Zhores A. Medvedev and Roy A. Medvedev, *A Question of Madness* (London: Macmillan, 1971).

42. Ruth Benedict, *Patterns of Culture* (Boston: Houghton Mifflin, 1934); Erwin H. Ackerknecht, "Psychopathology, Primitive Medicine and Primitive Culture," *Bulletin of the History of Medicine*, 14:30–67, 1943; Marvin K. Opler (ed.), *Culture and Mental Health* (New York: Macmillan, 1959); Robert B. Edgerton, "Conceptions of Psychosis in Four East African Societies," *American Anthropologist*, 68:408–424, 1966.

43. Zack Lynch, "Neuropolicy (2005–2035): Converging Technologies Enable Neurotechnology, Creating New Ethical Dilemmas," in William Sims Bainbridge and Mihail C. Roco (eds.), *Managing Nano-Bio-Info-Cogno Innovations: Converging Technologies in Society* (Berlin: Springer, 2006. pp. 173–191).

44. President's Council on Bioethics, *Beyond Therapy: Biotechnology and the Pursuit of Happiness* (Washington, DC: President's Council on Bioethics, 2003, p. 10).

45. President's Council on Bioethics, *Beyond Therapy: Biotechnology and the Pursuit of Happiness* (Washington, DC: President's Council on Bioethics, 2003, p. 13).

46. Wrye Sententia, "Neuroethical Considerations: Cognitive Liberty and Converging Technologies for Improving Human Cognition," in Mihail C. Roco and Carlo D. Montemagno (eds.), *The Coevolution of Human Potential and Converging Technologies* (New York: New York Academy of Sciences, 2004, p. 223).

47. William E. McAuliffe and Robert A. Gordon, "A Test of Lindesmith's Theory of Addiction: The Frequency of Euphoria Among Long-Term Addicts," *American Journal of Sociology*, 79(4):795–840, 1974.

48. Wrye Sententia, "Cognitive Enhancement and the Neuroethics of Memory Drugs," in William Sims Bainbridge and Mihail C. Roco (eds.), *Managing Nano-Bio-Info-Cogno Innovations: Converging Technologies in Society* (Berlin: Springer, 2006, pp. 153–171).

49. J. S. Albus, W. S. Bainbridge, J. Banfield, M. Dastoor, C. A. Murray, K. Carley, M. Hirshbein, T. Masciangioli, T. Miller, R. Norwood, R. Price, P. Rubin, J. Sargent, G. Strong, and W. A. Wallace, "Theme D Summary," in Mihail C. Roco and William Sims Bainbridge (eds.), *Converging Technologies for Improving Human Performance* (Dordrecht, Netherlands: Kluwer, 2003, p. 276).

50. Reginald G. Golledge, "Understanding Geographic Space Without the Use of Vision," online at http://www.ucalgary.ca/~rjacobso/haptic/research/NSFundexecsum.pdf

51. Reginald G. Golledge, "Spatial Cognition and Converging Technologies," in Mihail C. Roco and William Sims Bainbridge (eds.), *Converging Technologies for Improving Human Performance* (Dordrecht, Netherlands: Kluwer, 2003, pp. 122–140); "Multidisciplinary Opportunities and Challenges in NBIC," in Mihail C. Roco and Carlo D. Montemagno (eds.), *The Coevolution of Human Potential and Converging Technologies* (New York: New York Academy of Sciences, 2004, pp. 199–211).

52. Jack M. Loomis, "Sensory Replacement and Sensory Substitution: Overview and Prospects for the Future," in Mihail C. Roco and William Sims Bainbridge (eds.), *Converging Technologies for Improving Human Performance* (Dordrecht, Netherlands: Kluwer, 2003, pp. 213–224).

53. Britton Chance, "Interacting Brain," in Mihail C. Roco and William Sims Bainbridge (eds.), *Converging Technologies for Improving Human Performance* (Dordrecht, Netherlands: Kluwer, 2003, pp. 224–226); Edgar Garcia-Rill, "Focusing the Possibilities of Nanotechnology for Cognitive Evolution and Human Performance," in Mihail C. Roco and William Sims Bainbridge (eds.), *Converging Technologies for Improving Human Performance* (Dordrecht, Netherlands: Kluwer, 2003, pp. 227–232); Jeffrey M. Stanton, "Nanotechnology, Surveillance, and Society: Methodological Issues and Innovations for Social Research," in Mihail C. Roco and William Sims Bainbridge (eds.), *Nanotechnology: Societal Implications–Individual Perspectives* (Berlin: Springer, 2006, pp. 87–97).

54. Sherry Turkle, "Sociable Technologies: Enhancing Human Performance When the Computer Is Not a Tool but a Companion," in Mihail C. Roco and William Sims Bainbridge (eds.), *Converging Technologies for Improving Human Performance* (Dordrecht, Netherlands: Kluwer, 2003, pp. 150–158).

55. Mihail C. Roco and William Sims Bainbridge (eds.), *Converging Technologies for Improving Human Performance* (Dordrecht, Netherlands: Kluwer, 2003).

Chapter 6

Unification of Science

Science is becoming unified around a relatively small number of theoretical principles (such as conservation, indecision, configuration, interaction, variation, evolution, information, and cognition) and research methodologies (including new imaging technologies and computer simulation). To take advantage of unification, science and engineering education must be transformed, producing a new kind of field-spanning scientist, engineer, and technician. With cognitive science as a base, convergence will expand into the social sciences and humanities to an unknown extent, with unpredictable implications for humanistic values.

CREATING CONVERGERS

The idea that a new kind of scientist could be educated to work across fields is not new. I was first introduced to it as a child when I read the novel *Voyage of the Space Beagle*, by A. E. van Vogt.[1] To those who do not recognize the historical reference in the title, the name of this book must appear strange, even silly. But the *Beagle* was the ship that carried Charles Darwin around the world and enabled him to make his epochal discoveries about evolution. The *Space Beagle* is an interstellar ship that encounters strange forms of alien intelligence, challenging the hero, Dr. Elliott Grosvenor, to understand their behavior. It is widely believed that this novel served as the inspiration for the *Alien* series of movies, but the character of Grosvenor was left out of them—with the disastrous results that can plainly be seen by anyone who watches these films.

Grosvenor was a scientist of everything and of nothing. Like the Doctor on the long-running British television series *Dr. Who*, he cannot say exactly which field he is a doctor of. He is a *nexialist*, the opposite of a specialist. He follows arcane principles of abstraction and analogy to take information and principles from all sciences, thereby solving perplexing mysteries that no one

143

science can solve alone. At first, the crew of the *Space Beagle* is contemptuous of Grosvenor, considering him a "jack of all trades and master of none" in a civilization that prizes professional specialization. Then Grosvenor saves all of their lives by being able to understand and, therefore, to handle alien situations through nexialism.

Van Vogt, although a fiction writer, hoped that knowledge of the human mind really could achieve vast breakthroughs during his own lifetime, and he became an advocate for a mental discipline called *general semantics*.[2] This intellectual movement was founded by Alfred Korzybski in the period between the two world wars, and it drew upon ideas that were floating around in philosophy, linguistics, and related fields.[3] It held that language distorted thought as much as it facilitated thinking, and it taught various methods and supposed insights for transcending the cage of words in which language imprisoned untrained minds.

Like its more famous cousin psychoanalysis, general semantics was neither pure hokum nor pure science, and it had some positive influence on the academic community. For example, my sophomore English teacher at the Choate boarding school, Porter Dean Caesar, assigned the watered-down general semantics textbook *Language in Thought and Action*, which was written by Korzybski's disciple, S. I. Hayakawa (who later became a U.S. senator).[4] One of Korzybski's principles was "The word is not the thing," an insight warning us that we must not take the categories of language too seriously. In reaction to this principle, but also to push us to use wider vocabularies, my teacher would take 40 points off the grade of any paper or test in which his students' used the word "thing." Similarly, Korzybski argued for the development of a non-Aristotelian logic, one recognizing that declarative statements need not be either true or false.

In Chapter 5, I distinguished between two different approaches in artificial intelligence, structural–computational theory and neural networks. Korzybski's theories were far more compatible with the latter. My own early exposure to general semantics may explain why I feel much more comfortable with the (frankly) unclear and uncertain neural net approach—writing my first neural net programs in 1982 prior to their wave of popularity in the late 1980s—and less comfortable with the more classical and dominant artificial intelligence approach based on logical manipulation of hard-edged concepts. We should not give general semantics too much credit, because similar ideas were common in twentieth-century intellectual circles. Nevertheless, this theory reminds us that several generations of visionaries have hoped that new cognitive principles could be developed that would enable us to master a wider and deeper corpus of knowledge. Success in this profoundly difficult endeavor would be crucial for education as well as for the practice of scientific research.

The original report on the societal implications of nanotechnology included a prominent section on education, which immediately raises the issue of what one will teach to whom.[5] George Whitesides and J. Christopher Love of Harvard University showed that nanotechnology is itself a convergent field, bringing together elements from physics, chemistry, materials science, and information technology. As Whitesides and Love explain:

> Since the problems in nanoscience and technology will require a broad range of talents, collaborative research will probably be more the norm than the exception in universities, and perhaps in industry. Developing styles of research that span multiple disciplines is a growing skill in universities, but one whose importance needs constant reinforcement.[6]

If convergence demands much, however, it also offers much. For example, Stephen J. Fonash of the nanofabrication facility at Pennsylvania State University observed that the convergence creating the field of nanotechnology will stimulate employment opportunities—not only for a few highly trained scientists and engineers, but also for large numbers of well-trained technicians:

> Nanoscience and nanotechnology are the meeting ground of biology, engineering, medicine, physics, and chemistry. As a result, development and manufacturing at the nanoscale will be increasingly involved with components of all these pure and applied sciences. Because this field is becoming so broadly based and so broadly utilized, practitioners will no longer be limited to the microelectronics industry as they once were but, instead, will be free to choose jobs and follow career paths in a variety of fields such as biomedical applications, MEMs [micro-electro-mechanical systems], microfluidics, opto-electronics, information storage, pharmaceuticals, and, of course, microelectronics.[7]

How will expertise be distributed across the members of collaborative research and development teams? Two patterns should probably be combined. First, most members should be experts in one field, while having some familiarity with the other fields, and should master technical tools that operate across domains. Mathematics, sometimes called "the universal language," is among the most prominent tools applied across disciplines. As M. Gregory Forest argued in that first report, it has been fundamental to nanotechnology.[8] Aside from mathematics, scientists in different fields often have trouble talking with one another, not only because of the different subject matter covered by their respective fields, but also because of the historical developments of their socially separate fields. Two very different scientific customs could be functionally equivalent, alternate ways of performing the same functions, just as different words could signify the same thing in different languages. At present, each field of science or engineering has to some extent its own culture—that is, the particular concepts and terminology it uses—and a key challenge for convergers is how to translate between these cultures or to find the larger language of which they are dialects.

Second, some members of the team should be experts in synthesis across fields, comparable to van Vogt's nexialists. Michael Gorman (Figure 6–1), who contributed to fully three of the Converging Technologies conferences plus the most recent nanotechnology conference, suggests several helpful ideas:[9]

- Establishing trading zones—that is, collaborative environments where representatives of different disciplines can exchange knowledge and develop cooperative relationships

Figure 6–1 Michael E. Gorman with a telescope (courtesy of Wiebe E. Bijker, Maastricht University). Building on a background in the history, ethics, and teaching of science and engineering, this researcher has identified social strategies for achieving convergence across fields.

- Training some students to be interaction agents, with broad technical expertise plus backgrounds in social science, who can help more specialized scientists and engineers collaborate with one another

- Developing Creole-type languages that span fields, based on shared metaphors and mental models, to facilitate communication

- Teaching all professionals to practice moral imagination, which seeks to promote the greater good of humanity rather than narrow interests

Aside from stressing the importance of mathematics, how shall we teach intellectual convergence? At its 2002 meeting in Maryland, the American Association of Physics Teachers advocated a policy of "physics first," teaching physics to students early in their high school educations "to lay the foundation for more advanced high school courses in chemistry, biology, or physics."[10] Traditionally, biology is taught first, on the assumption it is closer to children's personal experience of life and, therefore, easier to understand. Traditional biology taught an incoherent mass of facts, with a huge vocabulary naming the structures in plants and animals, plus the taxonomy of species. As a consequence, it did little to prepare students for courses in other sciences. Placing physics first means that students will learn principles that can be applied to understand chemistry, which in turn links to biology via biochemistry.

An excellent example of physics-based convergence in high school teaching is the two-decades-old Magnet Program at Montgomery Blair High School in Silver Spring, Maryland, which takes in about 100 gifted boys and girls every year from across Montgomery County.[11] The first four semesters require four science courses in this order: physics, chemistry, Earth science, and biology. When my elder daughter took this program, I observed that the transitions from physics to chemistry actually involved assignments concerning electron orbits that took quantum theory into account.

This sequence of courses in specific sciences is but one of four curricular themes; the other three are mathematics, computer science, and interdisciplinary research and experimentation. The math courses explicitly offer tools that are useful in the sciences, while at the same time encouraging students to advance in mathematics itself. The math courses include Magnet Geometry, Magnet Precalculus, Magnet Functions, and Analysis I. The first set of computer science courses specifically focuses on how scientists use computers, and actual programming begins in a second course on algorithms and data structures. In both of the first two years, students take a research and experimentation laboratory course, Techniques for Problem Solving.

The junior and senior years offer a variety of advanced courses across the natural sciences: Analytical Chemistry, Astronomy, Cellular Physiology,

Genetics, Mathematical Physics, Marine Biology, Optics, Origins of Science, Physical Chemistry, Plate Tectonics and Oceanography, Quantum Physics, and Thermodynamics. Math electives include Analysis II, Applied Statistics, Complex Analysis, Discrete Math, and Linear Algebra. Computer electives include Advanced Applications, Analysis of Algorithms, Artificial Intelligence, Computational Methods, Graphics, Modeling and Simulation, Networking, Software Design, and 3D Graphics. Two standard junior or senior classes are explicitly convergent: Materials Science and Robotics.

Juniors are expected to take a special class in research design, so as to prepare them for undertaking a senior project. In the summer between their junior and senior years, most do research internships, working in local labs, such as those at the National Institutes of Health (which is situated in Montgomery County). My daughter did her internship at the University of Maryland, working alongside graduate students under the direction of Professor Douglas Oard of the College of Information Studies and the Institute for Advanced Computer Studies. Her project involved designing and programming a software module that would track the geographic movements of multilingual Holocaust survivors who had been interviewed for Steven Spielberg's Survivors of the Shoah Visual History Foundation. This endeavor combined her interest in computers with her interest in language; she has studied French, Latin, and Japanese, and was the editor of Blair's literary annual. In their senior year, the students write scientific reports, and one evening they actually give "poster" presentations like those encountered at a scientific conference.

The students in the Blair Magnet Program are unusually gifted, so it is natural to wonder whether the "physics first" approach and integration of education across the natural sciences could succeed with the wider population. The students in Blair's program are highly motivated, with many of them enduring long bus rides across the county every day and bypassing their local high schools. Part of the motivation may be intellectual, because their challenging courses are interesting, but there is also a sense that this program is extremely valuable for students' future careers. Many garner so many advanced placement credits that the universities they attend would allow them to graduate with a bachelor's degree in three years. In the case of the University of Maryland, there is an option to remain for four years but to earn a master's degree in that time.

One would never want to say what the limits of achievement are for any individual, but people who enter college without a good science and math background are unlikely to be very successful in technical careers. Both the students and their parents assume that graduates of the Blair Magnet Program will indeed become scientists, or perhaps engineers. Clearly, the nation needs more advanced technical professionals, so logically it should invest in more such programs.

To reach a much wider stratum of the population, science needs to develop more lucid ways of expressing its truths. The very innovations required for convergence at the professional level will help promote the scientific culture, because they offer concepts, language, and tools that enable bright people outside a particular field to understand its findings and methods, and to be able to accomplish something useful based on that knowledge. The widespread use of computers, as well as future technologies based on discoveries in cognitive science, will strengthen the ability of young people to handle material like that taught in the Blair Magnet Program. Although many parents today worry that video games will teach their children violence, in fact these exercises probably teach algorithms. That is, winning in many kinds of video games requires the player to learn planning, experimentation, and logical thinking through of alternatives. Given the instructive quality of new computer technologies, what we lack are the over-arching concepts to give people the cognitive tools with which to understand the complex world revealed by the sciences and created by the technologies.

EIGHT PRINCIPLES FOR CONVERGENCE

Eight high-level concepts can promote convergence by identifying analogies across fields of science and engineering: conservation, indecision, configuration, interaction, variation, evolution, information, and cognition (Figure 6–2). These principles are very high-level concepts, so they are not intended to be detailed or rigorous. Rather, each points in the general direction of a set of rigorous models, theories, processes, or formulae. To move forward in one domain D_1, we must look at how another domain D_2 has employed one or more of these concepts, and imagine how similar thinking could be employed in D_1. Through this exercise, domains become linked and begin to merge. Through convergence, they can serve the two main goals of science effectively.

One goal of science is *knowledge,* and from knowledge comes wisdom. Science should provide all educated people with an accurate, if qualitative, understanding of existence and human life. The over-arching principle is faith that the universe really is comprehensible. In the middle of the history of science, expertise requires fragmentation. Nevertheless, the end result of science should be a simple model that "comprehends" the many more complex models of subsidiary phenomena.

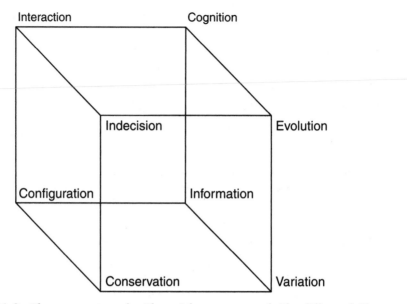

Figure 6–2 The convergence cube. These eight concepts can bridge different fields of science and technology, because singly or collectively they suggest theories and methodologies that apply to multiple domains of knowledge and human endeavor.

The second goal of science is to serve human material *welfare* through engineering design. To achieve higher-level functionality in engineering requires concepts for integrating systems across domains and scales. That is, converging technologies both require and promote the unification of science. The NBIC quartet demands a set of abstract but meaningful concepts that encompass many lower-level concepts and usefully integrate diverse fields. Other candidate concepts should be explored as well as the eight presented here; indeed, these eight are offered up merely as a starting point. Transformative concepts will offer both wisdom and power, if we can find the correct and effective way of thinking.

Conservation

Matter, energy, and possibly many other quantities are conserved: They are neither created nor destroyed, but merely transformed. Without a high degree of conservation, the universe would not persist or be stable enough for intelligent life to evolve. However, all conservation laws have limited scope. Some are only very approximate, such as the stability of crime rates in society, but nonetheless vary only within certain limits.

A classic question of philosophy and cosmology is stated as follows: "Why is there something rather than nothing?" Principles of conservation answer that question by reversing the question: "What suppresses chaos sufficiently to permit humans to perceive meaningful phenomena?" Without conservation, the cosmos would be a blooming, buzzing confusion. This is a version of the *anthropic theory*, holding that we perceive some features of the world owing to a selection effect. Only in a universe that had these features could complex biological organisms evolve, gain intelligence, invent science, and ask themselves why the universe possesses the features it does. Even in a conducive environment, biological evolution requires hundreds of millions of years, so the basic constituents of physical reality must hold a high degree of stability over that span of time, or there will never be any observers of any kind—let alone scientists.[12]

Whatever the merits of these ideas, the primary source of conservation is the stability of natural laws and parameters at the subatomic level, and many higher-level conservation laws can be reduced to conservation in physics. But it is also possible that conservation may emerge spontaneously at higher levels of complexity, through the operation of adaptive systems.

When governments first began collecting social statistics systematically, it became apparent that the rates of many kinds of human behavior were relatively stable from one year to the next, although different societies might have different characteristic rates. Some kind of homeostatic principle appears to operate in society, automatically adjusting rates of some individual behaviors to keep them close to the optimal level as part of a self-sustaining complex adaptive system. Perhaps the market is responsible: If there are too few doctors, members of this profession will be able to demand higher pay, which will attract more talented people to it. Or the political system may do the job: Increased crime rates will promote greater investment in police. Across the sciences, the concept of complex adaptive systems provides one possible explanation for empirical stabilities.

Indecision

In this context, *indecision* means undecidability, ambiguity of perception, or inability to predict. Undecidability, inconsistency, and uncertainty are not merely minor factors that inconvenience mathematicians and physicists, but may be a fundamental necessity for existence. If symmetry is a kind of conservation, then symmetry breaking is a kind of indecision. Without some initial inhomogeneity, perhaps generated by quantum effects in the very first moment of the "Big Bang," gravity would not have been able to concentrate matter into galaxies and stars. Indeed, estimating that inhomogeneity is a current challenge being targeted by astronomers.

If all parameters of existence were always conserved, then there could be no change, no events, and no life. Yes, if the universe were already a diverse collection of objects in motion, conservation alone would allow them to interact in complex ways. But the generation of the universe in the first place, and many kinds of small-scale events that occur throughout time, require unpredictable inputs. In physics, Heisenberg's uncertainty principle is the most familiar example, and Gödel's analysis of inconsistency and incompleteness is a prominent example in mathematics.[13] Words such as "indeterminacy" and "undecidability" are relevant here. So is "complementarity," which implies there is more than one framework in which to measure any phenomenon.

Configuration

The original inspiration for technological convergence is based on the unity of nature at the nanoscale, where atoms assemble into complex structures or configurations. Much of the secret of life resides in the precise structures of protein molecules. On a larger scale, social institutions are configurations, as are computer networks.

An excellent example is the way that a configuration concept from semiconductor physics found its way into sociology through the work of Harrison White, who has the distinction of possessing doctorates in both of these fields.[14] Electrons in a semiconductor can become dislodged from their atoms—a phenomenon below the nanoscale in which an electron may temporarily become sufficiently excited to leave the outer (valence) band of electrons in the particular atom. These runaway electrons may then move a few steps away from their original atoms—dynamics on the nanoscale—constituting an electric current. When an electron wanders away in this manner, it leaves behind a "hole" into which another electron may migrate, thereby leaving behind a hole at its prior location. Thus, as negative electrons move in one direction, positive holes travel the other way, and scientists working with semiconductors have found it useful to treat these holes as real objects.

Citing the 1950 book *Electrons and Holes in Semiconductors* by William Shockley, inventor of the transistor,[15] White applied this same idea to the movement of people and job openings in formal organizations that have relatively stable bureaucratic structures. For example, when the minister of a large church retires, his or her vacancy is likely to be filled by someone who held a position in a medium-sized church of the same denomination. This leaves a vacancy that can be filled by a minister moving up from a small congregation.

The mathematics of these occupational *vacancy chains* is comparable to the mathematics describing the movement of holes in semiconductors, so this convergence between physics and sociology is far more than just a shared metaphor.

Interaction

I take the concept of interaction from social psychology, but analogous concepts exist in many other disciplines. Elements of a system, such as individual people in a community, interact with one another in manifold ways. Two ideas especially inform our understanding of interaction: structure and emergence.

The structure of interaction in a system is a variant of configuration. Each element of the system has its own configuration and interacts with other elements in terms of the distinctive combination of their configurations. The clearest example is probably the interaction of chemical compounds, which is largely determined by the structures of the participating molecules. Emergence refers to the appearance of larger structures on the basis of smaller or simpler ones, such as in self-assembly (nanotechnology), evolution (biology), chunking (information science), or abstraction (cognitive science).

Structure and emergence are attributes of complex systems, in which the chaos of the constituent elements is connected in such a way that order reigns at the higher level of the system. Systems can be treated as entities that are elements in a larger system, like families within neighborhoods within cities within provinces within a nation, raising the question of whether the entire world is a system born out of interactions.

Variation

Although every proton is identical, physical objects containing protons differ greatly, even within species or categories. These differences can be conceptualized as different configurations of identical units. A standard mathematical way to conceptualize such variation is in terms of the probabilistic aggregation of indistinguishable events. Among the most widely used tool of statistics is the *normal distribution*, or *normal curve*, a way of conceptualizing and measuring variation when most cases are close to the mean and their dispersion around that mean is symmetrical and diminishing.[16]

Another common pattern of variation, which applies across many fields, is the Zipf distribution. It is named after George Kingsley Zipf, who believed this pattern reflected a fundamental natural principle of least effort.[17] Zipf's examples primarily concerned human language—for instance, an analysis of the frequency of different words in the novel *Ulysses* by James Joyce. After counting how many times a word was used (its frequency), the words were ranked in terms of their frequencies, with the number 1 being assigned to the most common word, 2 to the second-most common word, and so on. Zipf postulated that the product of each word's frequency and its rank was approximately constant across all words, both common and rare alike. Like the normal curve, Zipf's distribution can be used as an analytical tool, rather than assuming it must apply in all circumstances.

In support of his view that the underlying principle applied across many realms, Zipf cited the example of city size. Suppose we take the populations of all cities, towns and villages in a major country and arrange them in descending order of size, and then graph their populations. Graphed on log-log scales, this exercise produces an approximately straight line with a negative slope. The same distribution appears to fit the sizes of corporations, the sizes of religious congregations generated from computer simulations of social interaction, and nanoscale particles in magnetized ferrofluids.[18] One would predict that Zipf's distribution also roughly fits the sizes of solid bodies in the solar system (a few planets, many moons, huge numbers of asteroids, almost infinite numbers of nanoscale dust particles) and the populations of animal species on Earth (few elephants, many mice, vast swarms of gnats). Some debate has arisen regarding whether Zipf's discovery really reflects a meaningful natural law, or indeed what it means, but the distribution apparently qualifies as a pattern of variation or scaling law that is relevant to multiple sciences and technologies.[19]

Evolution

Natural selection from random mutation puts variation in the service of biological change.[20] Daniel Dennett has argued that evolution by natural selection can apply outside the biological realm, following these three principles:[21]

1. Variation: There is a continuing abundance of different elements.
2. Heredity or replication: The elements have the capacity to create copies or replicas of themselves.
3. Differential "fitness": The number of copies of an element that are created in a given time varies, depending on interactions between the features of that element and features of the environment in which it persists.

Some cosmologists believe that the pre-biological cosmos was also subject to natural selection in accordance with the anthropic principle, which resulted in a universe conducive to biology.[22] In Chapter 4, we showed how the biological principle of evolution by natural selection has been applied to culture, to computer programming, and to engineering design.

Information

Claude Shannon's classic information theory (data transmission theory), which emphasizes the concept of entropy, had immediate applications in cognitive science.[23] Indeed, theorists have long recognized its connection to thermodynamics.[24] Herbert A. Simon made the point in this way:[25]

> Information theory explains organized complexity in terms of the reduction of entropy (disorder) that is achieved when systems (organisms, for example) absorb energy from external sources and convert it to pattern or structure. In information theory, energy, information, and pattern all correspond to negative entropy.

Physicist John Archibald Wheeler coined the slogan "it from bit" to suggest that information is the primary stuff of the universe, rather than protons or energy.[26] Historian David Channell predicts, "In the future, we may come to see the second law of thermodynamics (entropy) as a consequence of information theory and not the other way around."[27] Any violation of a conservation law, or any indecision, entails the input of information. As Jim Spohrer of IBM says, "Change is information; information is change."[28]

Cognition

One of the most controversial ideas in the original *Converging Technologies* report was the notion that new technologies based on cognitive science should join nanotechnology, biotechnology, and information technology as a co-equal field. In Chapter 5, we noted that cognitive science is itself a convergence of multiple disciplines, including neurobiology (cognitive neuroscience), cognitive and perception psychology, aspects of linguistics, computer science (artificial intelligence), anthropology, and other fields. My own discipline, sociology, has been slow to join in the cognitive science synthesis, despite the fact that many of its theories and much of its research data are cognitive in nature.[29]

Many people conceptualize cognition as the epiphenomenal result of realities on the biological or chemical level, and thus as logically subordinate to the seven other principles discussed here. Alternatively, traditional modes of thought, such as those associated with the great religious movements that have survived from ancient times, tend to conceptualize all existence in terms of the cognition of God. I would suggest a different perspective—namely, considering the nature of reality as being inseparable from our perception of it. It would certainly be a profound error to say that reality is subjective, because this leads to magical thinking in which the world conforms to our wishes. But it is also wrong to say that the physical world exists in some objective sense, independent of anyone's ability to perceive it. There is no ground of being, analogous to the blackboard on which a divine scientist writes equations, across which the galaxies are arrayed. Rather, existence is the concrete relationship of objects and observers, relative to each other.

ETHICAL PRINCIPLES

The principles promoting convergence should not be merely intellectual tools capable of spanning traditional fields, but should also promote ethical development with the goal of enhancing human capabilities for universal benefit. In all the conferences on societal implications of nanotechnology and technological convergence, this point has been emphasized by philosophers, social scientists, and visionaries.

Where do human ethics and morality come from? One standard theory is that they are ultimately rooted in religion, and clearly the major religions of the world do assert ethical principles. Several chapters of my recent book, *Across the Secular Abyss*, have explored this issue.[30] The evidence that religion supports morality is rather weak, although under some specific circumstances good evidence indicates that religion can make a contribution. I have suggested earlier in this book that the unification of science, rendered more influential in regard to human culture by the success of technological convergence, erodes religious faith. Is the entire enterprise of converging technologies therefore a threat to morality?

Perhaps not, because religion's power to sustain morality is weak and weakening anyway. For example, within the United States, the strength of organized religion varies greatly across cities and states. So, too, do rates of crime. However, some crime rates correlate negatively with religion, whereas others do not. Apparently, religion has some capacity to deter minor offenses such as theft and illegal drug use, but has no power at all to deter serious offenses such as homicide. Studies of adolescents in different communities indicate that religious youths are less likely to be delinquent than irreligious youths, but only in religious communities. In secular areas, like much of the Pacific region, religious adolescents are just as likely to engage in delinquency as their irreligious counterparts. This suggests that the power of religion is not only weak but also contingent, depending on social support in the community for its strength. Other research I carried out and reported in *Across the Secular Abyss* indicates that the power of religion to deter suicide declined markedly during the twentieth century in the United States, and it may no longer be a significant factor in suicide prevention despite the apparent strength of faith.

Most studies of the connection between religion and morality have been conducted in the United States, in part because the United States is one of very few advanced industrial nations that has remained significantly religious. Poland is also religious, but it has been undergoing vast social and political changes, and we do not have a large, consistent body of research on the topic from that country. Cross-cultural studies, reported in *Across the Secular Abyss*, indicate that religion is more popular and has more power not only in poorly educated societies, but also in societies where many people experience unusual personal insecurity, and where social services cannot be relied upon to provide great help to people in distress. This finding suggests that religion will become not just weaker but less valuable as technological progress, economic development, and improvement of social institutions produce more secure and happier societies. The alternative, that revivalist religion will seek to halt scientific and technological progress, is addressed at length in *Across the Secular Abyss*.

It has long been obvious that religion is not the only source of morality, so morals might survive the decline of faith. Philosophers have attempted to derive ethical principles logically. Some, inspired by John Rawls, argue on the basis of deduction from first principles that humans should benefit equally from any important innovation.[31] In the second conference on societal implications of nanotechnology, Vivian Weil, Director of the Center for the Study of Ethics in the Professions at the Illinois Institute of Technology, commented:

> Some people anticipate that issues of justice and fairness may emerge in the distribution of benefits from nanotechnology developments. They warn against developing a counterpart to the digital divide in the nanotechnology area, producing a gap between those benefiting from the information technologies and those left out. The term "nano divide" has already come into use. A primary concern of justice is to avoid increasing inequalities in society. Here the concern is not to further disadvantage those already disadvantaged. By producing benefits inaccessible to those already disadvantaged, nanotechnology developments would be unjust.[32]

We shall encounter this view again in Chapter 7, but it is not widely held outside academic philosophy. An alternative view is that ethics arises in the interaction between individuals, who learn to cooperate through exchanges that benefit both parties, and who come to value their exchange partners and, therefore, act benevolently toward them.[33] From that perspective it is possible to derive a very different principle of distributive justice—namely, that people deserve to benefit in proportion to their contributions. In other words, the person who provides more value for society deserves to derive greater benefit from society.

This principle becomes problematic, however, the moment we ask what a given individual has actually contributed, who gets to decide what the value of that contribution might be, and how in calculating the value we should handle situations in which some people have been harmed even as others were benefiting from the contribution. Alan Ziegler contributed to the discussion of this last point at the fourth Converging Technologies conference, when he suggested we should not rush to judgment about possible indirect harm caused by convergence.

Although many social scientists have studied how norms arise in the give-and-take of social interaction, preeminent converger Edward O. Wilson and other sociobiologists have suggested that cooperation may be programmed into the genes of many species. Perhaps much of the altruism we find in normal families is the result of instinct shaped by the biological evolution of our species, in which children are heavily dependent upon their parents. That evolution must have taken place when humans lived in small hunter-gatherer bands, containing perhaps 40 individuals on average, and interacted occasionally with neighboring bands—sometimes in harmony and sometimes in hostility. These cooperative instincts likely shape behavior by triggering behavioral propensities in reaction to seeing sadness on the face of someone whom we know intimately or by some other cues connected to sharing our entire lives with the person.

If that is the case, then these instincts would not operate in many circumstances of modern life, as when a person buys and sells in a widely distributed market of strangers, or when an individual flies a bomb-laden aircraft far above the territory of another nation. The intimate cues that might otherwise trigger the benevolent instincts are absent in such cases. Religion, therefore, may have been a solution to the problem of how to enforce norms and support altruism outside the intimate small band or family. Much of human history can be seen as the struggle to build larger and larger social groups, with a host of new cultural innovations emerging in the attempt to do so. However, consideration of the bloody past of human history suggests that religion may not have been an especially good solution, nor one that may be able to survive development of a culturally diverse global civilization. In a scientific age, we need a convergent social science to help us establish appropriate ethical principles.

SOCIAL RELATIONS

It is often wise to become familiar with the past before planning the future. More than 50 years ago, a valiant series of attempts was made to create a unified social and behavioral science that would have been quite compatible with recent cognitive science. Those attempts largely failed, although their legacy lives on and could be revived in a new effort to achieve convergence. A major center of the apparently premature convergence of social science was at Harvard University. I came to know several of the key participants during my two periods at Harvard, first as a graduate student and then as an associate professor. In reality, though, the story begins before I was born.

In 1932, Lawrence Joseph Henderson was a professor of chemistry at Harvard, although his Wikipedia entry calls him a "physiologist, chemist, biologist, philosopher, and sociologist."[34] An early advocate of convergence, he had written two books about the remarkable coincidence that the universe—the domain of physics and astronomy—had exactly the right characteristics for the evolution of life—the topic of biology.[35] Recognizing that the social sciences were still in a primitive state of development, Henderson gathered around him a group of young Harvard men with active minds, in what came to be called the Pareto Circle because it took the works of Vilfredo Pareto as its starting point.[36] Pareto had sought to delineate the proper division of labor between economics and sociology in understanding human society, thus contributing as much to their convergence as to their separation.[37] Among the participants in the Pareto Circle were sociologists Talcott Parsons and George Caspar Homans, whom I knew and whose works I studied in graduate school 40 years later.

The Harvard sociology department was originally established in 1931, with Henderson as a part-time member, but in 1946 the university combined sociology with social anthropology, social psychology, and clinical psychology to form the convergent Department of Social Relations.[38] Talcott Parsons, who was the first chairman of this department, and several allies attempted to establish a particular theory of human social behavior as the dominant orthodoxy of social science, and they succeeded to a significant extent for about two decades.

This episode deserves mention here for two reasons. First, it illustrates the challenges of convergence, notably the danger that a false unification can be achieved on the basis of vague ideas and a dearth of scientific evidence. Second, the particular theory, *structural functionalism*, looks like a very good candidate for convergence with cognitive science and hence a potential bridge to bring the social sciences into an alliance with the natural sciences. The story also revolves around George Caspar Homans, my own primary teacher at Harvard and the chief opponent to Parsons' orthodoxy.

Many years later, I participated in a kind of retirement celebration for Talcott Parsons, who left Harvard in 1973, in the form of a lecture by Robert N. Bellah, a disciple of the great man. In the windowless basement lecture room of William James Hall, Bellah argued that Parsons was the pinnacle of sociological progress, heir to the major intellectual tradition in sociology, and someone whose roots lay in the classic work of the German, Max Weber, and the Frenchman, Emile Durkheim, that Parsons had largely introduced to American sociology. Bellah drew a diagram across the top of the long blackboard, starting (if memory serves) with St. Augustine, through Weber and Durkheim to Parsons. He also depicted a much weaker tradition from Thomas Hobbes to George Homans below it on the blackboard.

Students still read Hobbes' 1651 book *Leviathan*, a work of social and political philosophy that imagines human life would be chaotic and miserable were it not for our social agreements to cooperate.[39] Prior to the social contract that established law-abiding society, Hobbes famously said, life was "solitary, poore, nasty, brutish, and short." As I looked at Bellah's meager diagram from Hobbes to Homans, I could not help but think he had drawn it nasty, brutish, and short, clearly implying that Homans was heir to an inferior tradition that did not deserve much notice. I looked around, saw that Homans was not in the room, and weakly rose to his defense.

Parsons treated society as a unit, an entity that existed quite apart from the human beings who served as its components. Homans viewed society as the chaotic net result of the actions of individual humans, having no existence in and for itself. One of Parsons' sources, Emile Durkheim, had taken upon himself the reform of the French education system along lines inspired by socialist ideology. To do so, he thought he needed to establish sociology as an autonomous discipline, one that could not be reduced to psychology. In his remarkable book *Suicide*, Durkheim went so far as to argue that the apparently most solitary act, self-murder, was actually a feature of society as a super-individual entity. Another of Parsons' sources, Max Weber (most famous for *The Protestant Ethic and the Spirit of Capitalism*, which suggested that Protestant beliefs and values were somehow implicated in the rise of capitalism) emphasized the importance of socially shared concepts. Parsons synthesized Durkheim's and Weber's ideas with those of similar writers in his 1937 book, *The Structure of Social Action*.[40]

As Homans reports in his autobiography, when Parsons asked him to read the manuscript of *The Structure of Social Action*, he hated it but dared not say so, because Parsons was already a Harvard faculty member whereas Homans had not yet found academic employment. Privately, Homans thought the Parsons' book was devoid of scientific methodology or empirical data:

> It was another book of words about other persons' words. Rarely did it make contact with actual human behavior. In such a book it is easy to claim one has demonstrated whatever one wants to demonstrate. Social science bulges with books of this sort.[41]

For his part, Parsons believed he had discovered the basis of convergence in the social sciences through his examination of the thoughts of great thinkers who "had been converging on a single theoretical scheme."[42] Parsons also attempted to incorporate ideas from economics and biology, especially the concept of *system* and the idea that the parts of a system perform functions necessary for its survival. Despite Homans' skepticism of *The Structure of Social Action*, Lewis Coser recalls, "This work was a landmark in that it set a new course that was to dominate theoretical developments from the early 1940s until roughly the middle of the 1960s."[43]

In 1951, after Homans had joined the Social Relations faculty, Parsons attempted to get the department to adopt a similar book, *Toward a General Theory of Action*, as the Bible of sociology. Familiarly called the "Yellow Book," this collection of essays had been edited by Parsons and his colleague Edward Shils. Homans steadfastly opposed letting any single viewpoint define sociology, but structural functionalism did dominate the Social Relations department, and from there cast long shadows over U.S. sociology for many years thereafter.

The Yellow Book is almost unreadable, because it weaves a thick tapestry of abstractions based on undefined terms. As the perspective permeated sociology over time, its principles became much simpler: Every society has a culture that defines the norms people should follow and the values they should seek. The institutions of the society are collections of roles, and individuals are taught to play the roles the society needs for its survival. Because societies and roles differ, humans face complex choices between actions.

While recognizing that humans have instincts and desires that drive their actions, structural functionalism was not interested in them. Instead, it sought to categorize elements of thought and action in terms of cultural categories, most prominent among them the pattern variables. The pattern variables proposed by Parsons and Shils were five dimensions of variation, or dichotomies, that they believed could describe any norm, value, role, or action:

1. Affectivity – Affective neutrality
2. Self-orientation – Collectivity orientation
3. Universalism – Particularism
4. Ascription – Achievement
5. Specificity – Diffuseness[44]

Intense anger that your nation has been insulted would be affective and shows a collectivity orientation, for example; it is emotional rather than cool, and oriented toward your group rather than toward you as an individual. Particularism could describe a judge giving members of her own family an advantage in court, rather than following norms of universal fairness. The king enjoys ascribed status, whereas a sports star must achieve it. The fifth pattern variable could describe a relationship between two people, which could be very specific, like the one between a shopkeeper and a customer that carries very few expectations, versus the diffuse and multifaceted relationship between two lovers. On occasion, each of the five distinctions might be useful in understanding some social context, but Parsons and Shils outrageously claimed that these were the only dimensions that defined characteristics of social phenomena.

When I interviewed Homans for a Harvard newsletter in February 1987, he attempted to summarize the theoretical framework proposed by Parsons and his associates:

> I would describe it as a series of definitions of words locked into a very elaborate scheme of classification, and the scheme of classification was based on the notion of each cell in the classification representing a functional necessity for a society. But the real trouble with it was that it was a scheme, and you could fit all sorts of people and social behavior into it, but it never consisted of any propositions—that is, saying X varies as Y. I'd already made up my mind that you couldn't have any science without propositions of that general nature, and a classification scheme, though no doubt interesting, was worthless unless it led to these. The interesting thing is that I think that the so-called sociological theorists today, most of them, still haven't gotten this notion clearly in their minds.[45]

It is important to realize that structural functionalism was a cognitive theory, describing (whether accurately or not) what its authors believed took place within the human mind. At one point, Parsons and Shils distinguished *cognition* from *cathexis* and *evaluation*.[46] In truth, these were three different functions of the human mind: (1) how a person understands something, (2) how the person feels about it, and (3) how the person decides which of several actions to perform. Thus, accuracy aside, structural functionalism was potentially compatible with the cognitive science movement. Perhaps unfortunately, structural functionalism suffered a catastrophic loss of plausibility among sociologists in the 1960s, just before cognitive science was ready to take off.

The crucial flaw in structural functionalism was the functionalist assumption that the particular norms and institutions of a given culture are necessary for the survival of the society. As soon as the Yellow Book was published, critics complained it did not account for social change and too readily assumed that societies were integrated—when they were not busy complaining that the theory was too vague to be tested scientifically, that is.[47] In the rebellious 1960s, many people on university campuses questioned traditional norms and doubted the objectivity of standard institutions.[48] Two students of Parsons, Kingsley Davis and Wilbert Moore, had argued that society bestowed its highest rewards on those kinds of work most essential to its own survival, thereby creating the stratification system of social classes.[49] But Marxists and moderate sociologists alike came to feel that the system of social classes was unjust, rather than functionally necessary.

BEHAVIORAL SOCIAL SCIENCE

Although Homans opposed Parsons, as a fellow alumnus of the Pareto Circle he shared the same hope to create a unified social science. Whereas Parsons looked to European social theory, psychoanalysis, and cultural studies for inspiration, Homans looked beyond the Social Relations department to economics and behavioral psychology, notably to the work of his friend B. F. Skinner, the famous behaviorist in Harvard's Department of Psychology.[50] Homans argued that social science should be a single field, encompassing psychology, anthropology, sociology, economics, political science, and even history and linguistics.[51] Further, he believed that social science should become one of the natural sciences, closely connected through behavioral psychology to biology, and from there to chemistry, physics, and all the others. Thus Homans was a pioneer converger, even more than Parsons and his far-flung crew had been.

This position was a very controversial stance for a sociologist, because it violated Durkheim's principle that sociology must be an autonomous discipline that could not be diminished by reducing it to the principles of some other science. Homans, for his part, was quite happy to be a reductionist, deriving sociological principles from psychological ones, psychological principles from biological ones, and so on down to the fundamental laws of physics. He believed that theories in any science needed to be expressed as formal propositions, logically connected to each other, in which every concept is well defined and from which specific predictions can be deduced. Somewhat pessimistically, he admitted, "The issue for the social sciences is not whether we should be reductionists, but rather, if we were reductionists, whether we could find any propositions to reduce."[52]

Homans believed that scientific theories should be constructed formally, after the model of Greek geometry, in a series of statements (propositions) that begin with axioms and abstract descriptions of empirical conditions, and then logically derive predictions. In his popular class Fundamental Social Processes, and in the two editions of his textbook *Social Behavior*, he tried to show how a new social science could be created that would be both deductive and empirical. Having served as his chief teaching assistant, I can testify that his class lectures superficially seemed rather simple, enlivened by imitations of Skinner's angry pigeons and of Homans himself swimming in the ocean and trying to deduce propositions about the temperature of the water under various weather conditions. But for any student who was intellectually alert, it was profoundly challenging, because it demanded that we take seriously the idea that sociology should be a science, when so many practitioners wanted it to be the ideological wing of the socialist movement or a pleasant backwater of the humanities.

Certainly, Homans' axioms were quite simple. By the time he had finished polishing them, in 1974, they were as follows:[53]

1. For all actions taken by persons, the more often a particular action of a person is rewarded, the more likely the person is to perform that action.

2. If in the past the occurrence of a particular stimulus, or set of stimuli, has been the occasion on which a person's action has been rewarded, then the more similar the present stimuli are to the past one, the more likely the person is to perform the action, or some similar action, now.

3. The more valuable to a person is the result of his action, the more likely he is to perform the action.

4. The more often in the recent past a person has received a particular reward, the less valuable any further unit of that reward becomes for him.

5a. When a person's action does not receive the reward he expected, or receives punishment he did not expect, he will be angry; he becomes more likely to perform aggressive behavior, and the results of such behavior become more valuable to him.

5b. When a person's action receives reward he expected, especially a greater reward than he expected, or does not receive punishment he expected, he will be pleased; he becomes more likely to perform approving behavior, and the results of such behavior become more valuable to him.

The first axiom is one way to state Skinner's fundamental proposition of operant conditioning, that an intelligent organism will adjust its behavior to maximize the rewards it can obtain from the environment. This proposition is behaviorist because it refers only to things that can be observed; it says nothing about what happens inside the mind or brain of the organism.

The second axiom, however, merely pretends to be behaviorist; it is really cognitive. The key word—"similar"—appears twice. How does a person decide what is similar to what? That judgment is cognitive, shaped to some extent by the biological nature of our sensory organs, but also shaped by learning and by language.

The third axiom reveals its fundamentally cognitive nature in the word "valuable." Human instincts are only very approximate, and while generally motivated by the physiological sensations of pleasure and pain, in most cases we need to learn the values of actions or objects. This proposition opens the door to the socially defined systems of value that Parsons liked so much but Homans avoided.

Homans' fourth and fifth axioms introduce emotion, which is the twin of cognition. The fourth proposition draws on both biology and economics to recognize the importance of deprivation or satiation in temporarily adjusting the relative values of different rewards. Hold a hungry man's head under the water, and he will forget his hunger, for example, in an increased desire to breathe. The first half of the fifth proposition (5a) is based on the work of psychologists in the early 1940s about the connection between frustration and aggression.[54] The second half (5b) reflects the opposite tendency of people to express thanks and approval when things go especially well. If there can be any doubt that Homans' theory was fundamentally cognitive, it is dispelled by his enthusiastic analysis of the exchange of advice for approval between human beings, which is his prime example of social exchange in *Social Behavior*.[55] Advice concerns the communication of information, approval is a psychic reward, and both are largely cognitive in nature.

When the influential sociologist of religion, Rodney Stark, and I set out to develop a theory of religion following Homans' model, we began with a fresh set of 7 axioms and then sought to show how 344 other propositions could be derived from them, with the help of 104 definitions. Our axioms explicitly described key features of human cognition, and of the world in which humans dwell. Because our aim was to derive religious belief and behavior from human nature and the conditions of human life, we did not mention religion in the axioms, just as one would not have an axiom about the area of a triangle if one wanted to derive the formula for calculating that area from first principles. Chapter 5 briefly mentioned our chief line of argument, but the reader who seeks more details should consult our book, because here the key point is that Homans' approach, like Parsons' approach, leads inexorably to cognitive science.

Here are our seven axioms:[56]

1. Human perception and action take place through time, from the past into the future.

2. Humans seek what they perceive to be rewards and avoid what they perceive to be costs.

3. Rewards vary in kind, value, and generality.

4. Human action is directed by a complex but finite information-processing system that functions to identify problems and attempt solutions to them.

5. Some desired rewards are limited in supply, including some that simply do not exist.

6. Most rewards sought by humans are destroyed when they are used.

7. Individual and social attributes that determine power are unequally distributed among persons and groups in society.

Our first axiom may seem rather too cosmic for a sociological theory, but we quickly define past and future in terms of cognition and action. The past consists of the universe of conditions that can be known but not influenced. The *future* consists of the universe of conditions that can be influenced but not known. Once you know an event, it is no longer part of the future, so cognition is primarily based on memories of the past. The second axiom differs from Homans' axiom in that it recognizes that human perception and cognition are involved in deciding what is rewarding. The proposition that rewards vary in kind, value, and generality means that the pursuit of rewards is a complex business for humans, involving much work of different kinds to obtain different rewards.

The fourth axiom takes the bull by the horns, stating explicitly that human cognition is central to action. Three definitions expand on this axiom:

- The *mind* is the set of human functions that directs the action of a person.
- Human *problems* are recurrent situations that require investments (costs) of particular kinds to obtain rewards.
- To solve a problem means to imagine possible means of achieving the desired reward, to select the one with the greatest likelihood of success in light of the available information, and to direct action along the chosen line until the reward has been achieved.

What could be more cognitive than this? And yet our propositions were intended as a sociological theory (our book won the Outstanding Scholarship Award of the Pacific Sociological Association in 1993).

The fifth and sixth axioms introduce tragedy and toil into human life. Because many important rewards are often unobtainable (cure of a fatal illness, for example), people desperately seek hope that the limitations of the material world can somehow be transcended. That is the hook on which religion hangs. The fact that many rewards are what astronauts call *consumables* implies that people must repeatedly seek the same rewards, which they can often get from exchanges with other people, and those repeated exchanges build society.

For more than 20 years I have brooded about whether the seventh and final axiom is really necessary. It introduces inequality and politics into the human equation, but perhaps careful deductions from axioms 3 and 5 could accomplish the same thing.

It is nice to receive awards, laudatory reviews, and the supportive words of colleagues at social sciences conventions. It is nice to receive many invitations to publish without the irksome necessity of peer review. But the sorry fact is that no theoretical formulation, no matter how clear or sophisticated, has gained universal acceptance—or even reasonably wide acceptance—across the social sciences. In 1970, just months before I arrived at Harvard, the sociologists left the Department of Social Relations to establish again a separate sociology department. Two years later, while I watched in fascination, the psychologists and anthropologists who were still there departed for the existing departments of psychology and anthropology.

By the mid-1970s, sociology had become fragmented into a welter of specialties, each with its own limited theories and methods, with little to hold them together in a cohesive whole. Social psychology was split between psychology and sociology, with a roughly two-to-one ratio of practitioners, and researchers tended to fiddle with minor questions in their own small topic areas rather than seeking a grand synthesis.[57] Far from being ready to converge with other sciences, the core disciplines of the social sciences were splintering.

CONCLUSION

William H. Sewell has suggested that four factors ruined the social-science convergence that was attempted immediately after World War II: (1) convergence threatened the existing university departmental structure; (2) funding for multidisciplinary research was inadequate; (3) efforts to develop theoretical breakthroughs (like those described earlier) were unsuccessful; and (4) progress in research methodologies did not lead to major discoveries. Sewell's first two points suggest that a fresh effort could succeed today if significant resources were committed to this endeavor. As Chapter 7 will show, the world really needs a successful and objective social science, and the fragmentation of the past century has certainly not produced one. Sewell's third and fourth points are more troubling, because even a fragmented and impoverished social science should have been able to accomplish *something* in the past half-century, but it is not clear that it has.

Reflecting upon the disintegration of Harvard's Social Relations Department, Barry Johnston observes that its failure to develop a unified theoretical approach doomed social science to its current state of fragmentation, decline, and unscientific ideology.[58] I like to believe that a new social science can be built on the basis of cognitive science, perhaps a few years after the major NBIC convergence has progressed significantly. Until that happy day, how can we do without reliable social science?

Perhaps professional ethics can fill the moral gap, at least for deciding about technological development. Nanotechnologists and convergers can develop a set of principles and promulgate them through professional organizations. Such principles could be pragmatically valuable, supporting public trust and investment in new technologies, as well as upholding high values. After listening to the debates at all six major nanotechnology and convergence conferences, I can suggest a first draft in the form of four principles of ethical technological convergence:

1. Technical information about the possible directly harmful effects of an instance of technological convergence must be widely shared, so that a range of experts and stakeholders can participate in the debate about developing or deploying it.

2. To the extent possible, a convergent technology should be developed in the form most likely to limit its harmful effects while maximizing its benefits, including special efforts to mitigate harms when they cannot be entirely prevented.

3. Following the logic that incompetence is a form of unethical negligence among technical professionals, convergers must take account of the indirect effects of a new technology, such as socioeconomic disruptions.

4. Given that humans currently suffer many grievous hardships and dangers, the status quo is not inherently just or ethical, so with appropriate concern for other principles, innovative convergence is a positive good bordering on a necessity.

All four of these principles assume it is possible to understand the likely effects of a new technology, at least given some moderate effort to study the situation and likely impacts. To understand indirect and socioeconomic effects, we need a vigorous social science, which, sadly, we do not currently possess. Ideally, it should be possible to remake social science based on the basis of cognitive science supported by the three other convergent domains. To the extent that such an effort could succeed, we would be in a vastly improved position to understand how the unification of sciences and technologies could promote unification of human societies and provide the basis for a truly universal ethics.

REFERENCES

1. A. E. van Vogt, *The Voyage of the Space Beagle* (New York: Simon and Schuster, 1950).

2. A. E. van Vogt, *The World of Null-A* (New York: Simon and Schuster, 1948).

3. Alfred Korzybski, *Manhood of Humanity* (New York: E. P. Dutton, 1921); *Science and Sanity: An Introduction to Non-Aristotelian Systems and General Semantics* (Lakeville, CT: International Non-Aristotelian Library, 1948).

4. S. I. Hayakawa, *Language in Thought and Action* (London: G. Allen and Unwin, 1952).

5. Mihail C. Roco and William Sims Bainbridge (eds.), *Societal Implications of Nanoscience and Nanotechnology* (Dordrecht, Netherlands: Kluwer, 2001).

6. George M. Whitesides and J. Christopher Love, "Implications of Nanoscience for Knowledge and Understanding," in Mihail C. Roco and William Sims Bainbridge (eds.), *Societal Implications of Nanoscience and Nanotechnology* (Dordrecht, Netherlands: Kluwer, 2001, p. 136).

7. Stephen J. Fonash, "Implications of Nanotechnology for the Workforce," in Mihail C. Roco and William Sims Bainbridge (eds.), *Societal Implications of Nanoscience and Nanotechnology* (Dordrecht, Netherlands: Kluwer, 2001, pp. 179–180).

8. M. Gregory Forest, "Mathematical Challenges in Nanoscience and Nanotechnology: An Essay on Nanotechnology Implications," in Mihail C. Roco

and William Sims Bainbridge (eds.), *Societal Implications of Nanoscience and Nanotechnology* (Dordrecht, Netherlands: Kluwer, 2001, pp. 146–173).

9. Michael E. Gorman, "Expanding the Trading Zones for Convergent Technologies," in Mihail C. Roco and William Sims Bainbridge (eds.), *Converging Technologies for Improving Human Performance* (Dordrecht, Netherlands: Kluwer, 2003, pp. 424–428); Michael E. Gorman, "Collaborating on Convergent Technologies," in Mihail C. Roco and Carlo D. Montemagno (eds.), *The Coevolution of Human Potential and Converging Technologies* (New York: New York Academy of Sciences, pp. 150–177); Michael E. Gorman and James Groves, "Collaboration on Converging Technologies: Education and Practice," in William Sims Bainbridge and Mihail C. Roco (eds.), *Managing Nano-Bio-Info-Cogno Innovations: Converging Technologies in Society* (Berlin: Springer, 2006, pp. 71–87); Michael E. Gorman and James Groves, "Training Students to Be Interactional Experts," in Mihail C. Roco and William Sims Bainbridge (eds.), *Nanotechnology: Societal Implications—Individual Perspectives* (Berlin: Springer, 2006, pp. 301–305).

10. http://www.aapt.org/Policy/physicsfirst.cfm

11. http://www.mbhs.edu/departments/magnet/

12. John D. Barrow and Frank J. Tipler, *The Anthropic Cosmological Principle* (New York: Oxford University Press, 1986); William Sims Bainbridge, "The Omicron Point: Sociological Application of the Anthropic Theory," in Raymond A. Eve, Sara Horsfall, and Mary E. Lee (eds.), *Chaos and Complexity in Sociology* (Thousand Oaks, CA: Sage, 1997, pp. 91–101).

13. Werner Heisenberg, *Physics and Philosophy: The Revolution in Modern Science* (New York: Harper and Row, 1958); Jean Van Heijenoort (ed.), *From Frege to Gödel: A Source Book in Mathematical Logic, 1879–1931* (Cambridge, MA: Harvard University Press, 1967); Ernest Nagel and James R. Newman, *Gödel's Proof* (New York: New York University Press, 1958).

14. Harrison C. White, *Chains of Opportunity* (Cambridge, MA: Harvard University Press, 1970).

15. William Shockley, *Electrons and Holes in Semiconductors, with Applications to Transistor Electronics* (New York: Van Nostrand, 1950).

16. Glenn Shafer, "The Unity and Diversity of Probability," *Statistical Science*, 5(4):435–444, 1990.

17. George Kingsley Zipf, *Human Behavior and the Principle of Least Effort* (Cambridge, MA: Addison-Wesley, 1949).

18. Robert L. Axtell, "Zipf Distribution of U.S. Firm Sizes," *Science*, 293:1818–1820, 2001; William Sims Bainbridge, *God from the Machine* (Walnut Grove, CA: AltaMira, 2006); A. T. Skjeltorp, K. L. Kristiansen, G. Helgesen, R. Toussaint, E. G. Flekkoy, and J. Cernak, "Self-assembly and Dynamics of Magnetic Holes," in A. T. Skjeltorp and A. V. Belushkin (eds.), *Forces, Growth and Form in Soft Condensed Matter: At the Interface between Physics and Biology* (Dordrecht, Netherlands: Kluwer, 2004, pp. 165–179).

19. John L. Casti, "Bell Curves and Monkey Languages," *Complexity*, 1(1):12–15, 1995; Richard K. Belew, "Weighting and Matching against Indices," in *Finding Out About: A Cognitive Perspective on Search Engine Technology and the WWW* (New York: Cambridge University Press, 2000, pp. 60–104).

20. Stephen Jay Gould, "Trends as Changes in Variance," *Journal of Paleontology*, 62(3):319–329, 1988.

21. Daniel C. Dennett, *Darwin's Dangerous Idea: Evolution and the Meanings of Life* (New York: Simon and Schuster, 1995, p. 343).

22. John Leslie, "Anthropic Principle, World Ensemble, Design," *American Philosophical Quarterly*, 19(2):141–151, 1982; Andrei Linde, "The Self-Reproducing Inflationary Universe," *Scientific American*, 271:48–55, 1994.

23. Claude E. Shannon, "A Mathematical Theory of Communication," *Bell System Technical Journal*, 27:379–423, 623–656, 1948.

24. Steven C. Seow, "Information Theoretical Models of HCI: A Comparison of the Hick–Hyman Law and Fitts' Law," *Human–Computer Interaction*, 20:315–352, 2005.

25. Herbert A. Simon, *The Sciences of the Artificial* (Cambridge, MA: MIT Press, 1996, p. 172).

26. John Archibald Wheeler, *At Home in the Universe* (Woodbury, NY: American Institute of Physics, 1994).

27. David F. Channell, "The Computer at Nature's Core," *Wired*, 12(2), 2004, online at http://wired-vig.wired.com/wired/archive/12.02/view.html?pg=2

28. Personal communication.

29. Paul DiMaggio, "Culture and Cognition," *Annual Review of Sociology*, 23:263–287, 1997.

30. William Sims Bainbridge, *Across the Secular Abyss: From Faith to Wisdom* (Lanham, MD: Lexington, 2007).

31. John Rawls, *A Theory of Justice* (Cambridge, MA: Belknap Press of Harvard University Press, 1971).

32. Vivian Weil, "Ethics and Nano: A Survey," in Mihail C. Roco and William Sims Bainbridge (eds.), *Nanotechnology: Societal Implications—Individual Perspectives* (Berlin: Springer, 2006, pp. 172–182).

33. George Caspar Homans, *Social Behavior: Its Elementary Forms* (New York: Harcourt, Brace Jovanovich, 1974); Robert Axelrod, *The Evolution of Cooperation* (New York: Basic Books, 1984); David Gauthier, *Morals by Agreement* (New York: Oxford University Press, 1986).

34. http://en.wikipedia.org/wiki/Lawrence_Joseph_Henderson, retrieved August 27, 2006.

35. Lawrence Joseph Henderson, *The Fitness of the Environment* (New York: Macmillan, 1913); *The Order of Nature* (Cambridge, MA: Harvard University Press, 1917).

36. Barbara S. Heyl, "The Harvard 'Pareto Circle,'" *Journal of the History of the Behavioral Sciences*, 4:316–334, 1968.

37. George C. Homans and Charles P. Curtis, Jr., *An Introduction to Pareto: His Sociology* (New York: Knopf, 1934).

38. Suzanne Washington, "Towards a History of the Department of Sociology," *Sociology Lives*, 18(2):4–5, 2004.

39. Thomas Hobbes, *Leviathan* (London: Cooke, 1651), online at http://www.gutenberg.org/dirs/etext02/lvthn10.txt

40. Talcott Parsons, *The Structure of Social Action* (New York: McGraw-Hill, 1937).

41. George Caspar Homans, *Coming to My Senses: The Autobiography of a Sociologist* (New Brunswick, NJ: Transaction, 1985, p. 323).

42. Talcott Parsons, "Some Comments on the State of the General Theory of Action," *American Sociological Review*, 18(6):618–631, 1953.

43. Lewis A. Coser, "Sociological Theory from the Chicago Dominance to 1965," *Annual Review of Sociology*, 2:147, 1976.

44. Talcott Parsons and Edward A. Shils (eds.), *Toward a General Theory of Action* (Cambridge, MA: Harvard University Press, 1951, p. 77).

45. Quoted by William Sims Bainbridge in "An Interview with Professor Emeritus George C. Homans," *Sociology Lives* (newsletter of the Harvard University Department of Sociology), 3(3):6, 1987.

46. Talcott Parsons and Edward A. Shils (eds.), *Toward a General Theory of Action* (Cambridge, MA: Harvard University Press, 1951, p. 59).

47. W. W. Rostow, "Toward a General Theory of Action," *World Politics*, 5(4):530–554, 1953; Clarence Schrag, "Toward a General Theory of Action," *American Sociological Review*, 17(2):247–249, 1952.

48. Seymour Martin Lipset, *Rebellion in the University* (Boston: Little, Brown, 1971).

49. Kingsley Davis and Wilbert Moore, "Some Principles of Stratification," *American Sociological Review*, 10:242–249, 1945.

50. B. F. Skinner, *The Behavior of Organisms: An Experimental Analysis* (New York: Appleton-Century, 1938).

51. George Caspar Homans, *The Nature of Social Science* (New York: Harcourt, Brace and World, 1967, p. 3).

52. George Caspar Homans, *The Nature of Social Science* (New York: Harcourt, Brace and World, 1967, p. 86).

53. George Caspar Homans, *Social Behavior: Its Elementary Forms* (New York: Harcourt, Brace, Jovanovich, 1974, pp. 16, 22, 25, 29, 37, 39).

54. John Dollard, Neal E. Miller, Leonard W. Doob, O. H. Mowrer, and Robert R. Sears, *Frustration and Aggression* (London: K. Paul, Trench, Trubner, 1944).

55. George Caspar Homans, *Social Behavior: Its Elementary Forms* (New York, Harcourt, Brace, Jovanovich, 1974, pp. 53–57).

56. Rodney Stark and William Sims Bainbridge, *A Theory of Religion* (New York: Lang, 1987, pp. 27–32, 325).

57. Cookie White Stephan and Walter G. Stephan, *Two Social Psychologies* (Homewood, IL: Dorsey Press, 1985).

58. Barry V. Johnston, "The Contemporary Crisis and the Social Relations Department at Harvard: A Case Study in Hegemony and Disintegration," *American Sociologist*, 29:26–42, 1998.

Chapter 7

Unity in Diversity

Scientific and technological convergence could have the paradoxical effect of creating a unified world technical culture on which a great diversity of subcultures could be built. How might a new generation of social sciences (based on convergence with cognitive and information sciences) manage conflict in a world of diversity? Or will the world enter the future in ignorance of its meaning and incapable of responding to its challenges? This chapter draws on the ideas of critics of technology, ordinary citizens, social scientists, and those involved in the converging technologies movement. The fundamental but unanswerable question posed here is whether the convergence of all sciences and technologies could support the convergence of all human societies, or lay the groundwork for unity in diversity.

CRITICS OF CONVERGENCE

Many economists and other social scientists have long believed that the Industrial Revolution of a quarter millennium ago, which itself built on the great age of geographic discovery that occurred a quarter millennium before, is producing a unified, global technical and economic system. Such commercial globalization, many of them argue, will lead to cultural unity as well. When economists talk about "convergence," they mean the equalization of per capita income across nations they believe would follow globalization. Sociologists imagined that industrialization would cause a general convergence of values, beliefs, and even personality styles they called *modernity*.[1] In the 1990s, persistent doubts grew among social scientists, and the power of industrialization to achieve political and cultural convergence came under sharper scrutiny.[2] Recent social science review articles on industrialization and globalization describe multisided debates with no clear winners.[3]

It may be that a particular cultural constellation is required to develop the technologies on which industrialization and the information society are

175

based. This constellation may well involve secularization, democracy, and free-market economies. Once the technologies exist, however, they can readily diffuse to other societies that have totally different characteristics. In other words, technology may require particular social conditions but not cause them, so long as technology remains aloof from other institutions of the society. Perhaps a more thorough convergence, including the cognitive and social sciences that address political and cultural questions directly, would have more power to unite the world.

In the absence of an advanced, objective, and convincing social science, many observers of technological convergence are forced to fall back on pre-scientific ideologies in deciding how to respond to it. Outside the United States, an influential ideology of technology is based in the wreckage of Marxism and the concerns of moderate left-wing parties about globalization. It has become the perspective of a diffuse international movement that distrusts market economies and worries that any new technology would prove disadvantageous to workers and poor countries. This perspective does not come with a label attached, such as "Green," "Marxist," or "Luddite," yet is reminiscent of those ideologies—recalling that the historical Luddites of early nineteenth-century England attacked new technology only because they had no other way of defending workers' rights in a period of falling wages and increasing unemployment.[4] The perspective defines social justice largely in terms of equality, rather than believing that the rich and powerful have earned their positions. A corollary to this definition predicts that new technologies will give added power to the ruling classes, thereby increasing injustice.

Among the first carefully constructed critiques of nanoconvergence was the report *The Big Down*, published by an organization in Ottawa, Canada, calling itself the "ETC group." A brief primer on the background of the group will help place this critique in context. ETC is a successor to the Rural Advancement Foundation International (RAFI) and the heir to a long history of reform efforts dedicated to improving the conditions of poor farmers, in both advanced and developing countries. The Wayback Machine of the Internet Archive[5] has preserved RAFI's main webpages since May 29, 1998, when it described itself as follows: "RAFI is dedicated to the conservation and sustainable improvement of agricultural biodiversity, and to the socially responsible development of technologies useful to rural societies. RAFI is concerned about the loss of genetic diversity—especially in agriculture—and about the impact of intellectual property rights on agriculture and world food security."

In 2001, RAFI held a contest to select a new name, and ETC was the winner. The name "ETC" is an acronym for three topics: erosion, technology, and concentration. *Erosion* refers to trends toward environmental degradation, loss of species diversity, and declining knowledge and human rights. *Technology* refers to the danger that technological innovation will exacerbate economic inequalities if it takes place within an already unjust social system. *Concentration* refers to the growth of power by high-technology transnational corporations and other forms of oppression. These issues are manifestly those espoused by the international Green movement, and political conservatives would remind us that "behind the Greens stand the Reds."

Although the political origins of a thoughtful critique of nanoconvergence are far less important than its content, the ideological location of any particular critique does shape its perspective. The title of the ETC group's report, *The Big Down*, has a double meaning: Science and technology are going *down* into the nanoscale, and the result will be a depressing *downer*. The novelist Edgar Rice Burroughs used to joke that humans have "descended" from the apes in the sense that they have declined from the natural virtues of animals to the corruption of civilization—down the evolutionary drain, so to speak. The ETC report starts with the premise that NBIC will be a tremendously powerful technological revolution:

> Industry and governments promise that the manipulation of matter on the scale of the nanometer (one-billionth of a meter) will deliver wondrous benefits. All matter—living and non-living—originates at the nano-scale. The impacts of technologies controlling this realm cannot be overestimated: Control of nano-scale matter is the control of nature's elements (the atoms and molecules that are the building blocks of everything). Biotech (the manipulation of genes), Informatics (the electronic management of information), Cognitive Sciences (the exploration and manipulation of the mind), and Nanotech (the manipulation of elements) will converge to transform both living and non-living matter. When GMOs (genetically modified organisms) meet Atomically Modified Matter, life and living will never be the same.[6]

This power, ETC worries, will be unleashed without proper care and applied to goals that may restrict rather than enhance freedom. That is, ETC seems to have two categories of concerns. First, on a purely material level, ETC cites physical dangers to human health from nanoconvergence. Nanoparticles may cause illnesses, for example, and some of the new convergent technologies may have negative unintended consequences. In the "gray goo" scenario, self-reproducing nanoscale robots may eradicate life on our planet. Alternatively, in the "green goo" scenario, genetically altered microorganisms, perhaps united with nanotechnology components, threaten to destroy us.

This set of practical concerns becomes political, because ETC does not trust the governments and capitalist ruling classes it believes are promoting NBIC to pay proper attention to the welfare of the people.

Second, on a political level, ETC fears that the ruling classes will use the enhanced power they gain from NBIC to oppress the common people even further. *The Big Down* claims that the poor and marginalized members of society always suffer when governments promote technological innovation. It further claims that the United States is especially guilty of this form of inhumanity. Missing from ETC's argument, however, is scientific evidence or systematic analysis to back up these claims. Perhaps its best defense against this criticism is the claim that established agencies of government, which might be expected to possess the needed expertise, have failed to address these issues:

> The impact of converging technologies at the nano-scale is either unknown or underestimated in intergovernmental fora. Since nano-scale technologies will be applied in all sectors, no agency is taking the lead. Governments and civil society organizations (CSOs) should establish an International Convention for the Evaluation of New Technologies (ICENT), including mechanisms to monitor technology development.[7]

This argument seems a bit odd, given that issues such as these have, in fact, been addressed previously (e.g., in the original *Societal Implications* and NBIC reports) and that the U.S. NNI has gone to great lengths to include research on societal implications. Given the difficulty the world community has in reaching a consensus about almost anything, it seems a stretch to imagine that it could create an international agency to evaluate new technologies and presumably ban those deemed unsuitable. As yet, no official government funding initiative has been put forth to support converging technologies explicitly, so there cannot yet be a component devoted to societal implications of NBIC. However, our series of book-length Converging Technologies conference reports and the latest nanotechnology report continue to address ethical, legal, and social aspects of convergence, with increasing sophistication.

Given that *The Big Down* was a relatively early report (published in 2003), it is worth checking whether a more recent report in the same tradition has been able to assimilate the excellent work done by the participants in our six conferences. In 2006, Toby Shelley, who had previously written a book on poverty and the oil industry, published *Nanotechnology: New Promises, New Dangers* as part of a social issues series that is critical of the global capitalist system.[8] As a cultural document produced by a social movement, this book has a serious limitation that the movement itself shares—namely, the lack of an explicit intellectual tradition of scholarship. For all its faults, Marxism had an intellectual tradition, expressed in a vast library of publications that developed the ideas in some depth. *Nanotechnology* does not place its arguments in a well-defined theoretical context, although it might fit in the neo-Marxian analysis called "world systems theory" by left-wing sociologists.[9]

The movement's initial image of nanotechnology was the one promulgated by Eric Drexler, including the notion that self-reproducing nanoscale robots would soon take over many production tasks and could ultimately threaten human life. Refreshingly, Shelley's book acknowledges that Drexler's technical conception of nanotechnology was probably wrong. Like Drexler's own writings, however, *Nanotechnology* does not recognize that the notion of rampaging nanobots hearkens directly back to the original use of the word "robot" in the Czech play *R. U. R.* by Karel Čapek, in which robots represented the dehumanized working class revolting against the human ruling class. Although nanobots may be technically impossible, they are ideologically potent as a fresh metaphor both for the class struggle and for the millenarian hopes that some utopians have invested in nanotechnology.

On a somewhat more realistic level, Shelley's book expresses concern that nanotechnology may cause unemployment both among industrial workers and among producers of some raw materials in developing countries. While the imaginary nanobots might logically cause mass unemployment, at least among less skilled workers, Shelley believes that more modest nanotechnologies could also overturn the labor theory of value (which is the fundamental idea underlying Marxist economics). At the same time, his book expresses hope that nanotechnology could benefit humanity greatly, if somehow it could be rescued from the (supposedly) pathological forces of capitalism.

In one chapter titled "Nanotechnology Goes to War," and in another chapter about *nanowars* and domestic spying, Shelley notes the efforts to develop nano-enabled sensors for detecting chemical, biological, and radiological weapons, and discusses the importance of information for defense and the control of a nation's own citizenry. While discounting the strategic value of nanobot weapons, he does imagine that some forms of "mass murder" nano-weaponry could be developed. To the extent that nanotechnology will have any military implications, Shelley is concerned that its development will strengthen the global position of the United States. While not directly accusing this country of imperialism or fascism, as many left-leaning acolytes might do, he does express a lack of confidence that U.S. "military exploits" will benefit the people of the world.

A chapter on technological convergence notes (correctly) that much of the societal influence of nanotechnology is likely to result from its collaboration with other technologies. Despite the vast scope of technological convergence, this chapter is remarkably myopic, not even citing the original report, *Converging Technologies for Improving Human Performance.* Shelley's footnotes cite journalistic reports about new nanotechnology developments and position papers written by other members of his social movement, rather than reliable technical publications. If I might hazard a guess, I would speculate that the movement to which ETC and other groups cited by Shelley belong is not really interested in nanotechnology or convergence. Rather, it would like to promote a general vision of global justice that is politically out of fashion, and it is hunting for issues that might possibly increase public sympathy for its ideas.

LOOKING FORWARD

The ultimate impact of nanoconvergence depends on much more than simply the technical capabilities of the technology, the decisions that people make in implementing it, and the relevant government investments and regulations. That is, the effects of this movement will also reflect the ways in which existing social, economic, and cultural trends interact with these factors. The question is not how nanotechnology may change a stable world, but rather how the development of nanoconvergence will play into the forces that already swirl in an unstable and often chaotic world.

To answer this question, it will be important to identify current trends, understanding them both as forces that shape our daily lives today and as transformations that will lead to a radically different way of life in the future, even without nanotechnology. Indeed, we may find that nanoconvergence could be a stabilizing influence, rather than necessarily a destabilizing one. Nanotechnology may not be directly relevant for some major trends, although it may have powerful indirect implications through its closer connection to other trends.

There is much room to debate the nature and meaning of current trends, and each may be defined and described in different ways. In listing trends here, I draw upon not only my own reading of social science, but also the perspectives of about 20,000 people who responded to a question I placed in a major online questionnaire study, *Survey2000*, sponsored by the National Geographic Society. My question on this survey (which was the predecessor to *Survey2001* described in Chapter 2) read as follows: "Imagine the future and try to predict how the world will change over the next century. Think about everyday life as well as major changes in society, culture, and technology." Respondents were given a text area on the webpage in which to record their thoughts, anything from a single word to several paragraphs.

To assemble the respondents' ideas into a coherent whole, I went carefully through the text, copying out distinct thoughts about the future. The method of analysis was one I have used many times before—for example, in surveys about the possible goals of the space program for a book I published in the early 1990s.[10] Naturally, some ideas were expressed in roughly the same language by many different respondents to *Survey2000*, so I copied only the first or best expressions I encountered in such cases. This very time-consuming process eventually gave me a new file with slightly more than 5,000 text extracts. I then worked carefully through these 5,000 excerpts, combining and editing them into 2,000 clear statements of single ideas. Some are simple declarative statements asserting that something will be true in the year 2100. Others are more complex, suggesting what will cause a particular outcome, or combining factors to describe a general condition in the future. One by-product was a software module published on a CD-ROM disk that lets the user express his or her own evaluations of the ideas. Several articles or chapters based on these data have already been published or are in press.[11]

Naturally, the thousands of people who contributed ideas and evaluations were not of one mind. Indeed, they expressed several competing scenarios of the future for each of the major realms of human life. The following subsections describe their impressions of the future in four major realms: family and reproduction, culture and personality, societal institutions, and science and technology. These brief statements summarize some of the opinions expressed by the respondents, illuminated by social science research, and connected to nanoconvergence by theory. Ordinary people are not prophets, but their range of ideas does suggest the range of futures they are prepared to create.

Family and Reproduction

Survey2000 respondents held a wide variety of opinions about the future of the family. Some believed that the average family will become stronger over the coming century, with lower divorce rates and possibly a return to traditional family values. Others predicted that the long-term trend of rising divorce rates will continue, with dire consequences for children and for society as a whole. Others imagined that the family will be reinvented, with people living in a variety of quite different but equally viable family forms. To social scientists, it is not at all clear what could reverse the trends of the twentieth century, and in recent years the greatest concern has become the demographic collapse occurring in technologically advanced nations.

Most technologically advanced nations are headed toward or have actually entered population decline, with the notable exception of the United States. A useful source of demographic information is the online *World Factbook* maintained by the U.S. Central Intelligence Agency.[12] In August 2006, for example, it reported that the European Union had a population of about 456,950,000 the previous month, compared with 298,440,000 for the United States. Every year, 101 people die for every 100 who are born in the European Union, whereas only 58 die per 100 born in the United States. The EU population is still growing very slightly at an annual rate of 0.15% because of immigration; nevertheless, the imbalance between the birth and death rates will worsen as the age distribution of Europeans shifts to produce an older population. A constant population requires a fertility rate of about 2.1 to offset the facts that some girls die young and that more boys than girls are born (about 105 boys worldwide per 100 girls). The total fertility rate in the European Union is only 1.47, compared with a much healthier 2.09 in the United States.

The *World Factbook* offers fertility data about 114 places with populations of more than 5 million. Hong Kong has the lowest fertility rate, just 0.95, though in truth it should be considered an urban part of China. Eighteen countries have fertility rates of less than 1.5: Ukraine (1.17), Czech Republic (1.21), Poland (1.25), South Korea (1.27), Spain (1.28), Italy (1.28), Russia (1.28), Hungary (1.32), Slovakia (1.33), Greece (1.34), Austria (1.36), Romania (1.37), Bulgaria (1.38), Germany (1.39), Japan (1.40), Switzerland (1.43), Belarus (1.43), and Portugal (1.47). Note that this list includes some relatively poor countries, two Asian nations, and several rich European countries. Twenty-four nations have fertility rates between 1.5 and 2.0, including eight with populations exceeding 60 million: Thailand (1.64), United Kingdom (1.66), China (1.73), Iran (1.80), France (1.84), Vietnam (1.91), Brazil (1.91), and Turkey (1.92). Perhaps ironically, given the two countries' mutual hostility, the fertility rate of the United States (2.09) is closest to that of North Korea (2.10).

Poor nations need fertility rates higher than 2.10 to sustain their populations, because their mortality rates are high. Six nations in this group have populations of more than 100 million plus high fertility: Indonesia (2.40), Mexico (2.42), India (2.73), Bangladesh (3.11), Pakistan (4.00), and Nigeria (5.49). The entire world, with its population of about 6.5 billion, has a fertility rate of 2.59 children born per woman who completes her childbearing years. Clearly, humanity as a whole continues to increase, but the declines in advanced nations carry an implicit threat: The entire world may suffer population collapse if all societies become wealthy, well educated, and democratic.

For two decades, demographers have been aware that the fertility rates for technologically advanced nations might drop significantly below their replacement levels.[13] A nation that declines in population is likely also to decline in world influence.[14] Its market for goods and services may shrink, stifling its domestic economic development. A larger proportion of its population will be elderly and dependent; a smaller proportion will be young and creative. We may well wonder how the demographic trends will play out as population explosion continues in underdeveloped societies while fertility collapse threatens the long-term viability of developed societies.

The mutual implications of nanoconvergence and family trends for each other are not clear. Clearly, nano-bio convergence could play major roles in facilitating medical control of fertility (whether reducing or increasing the birth rate) and in increasing the life span, which has complex implications for how older people fit into families and communities. In principle, nano-enabled medical treatments could allow adults to produce children progressively later in an ever-increasing life span, thereby sustaining fertility at a level that keeps industrial society viable. Young adults could bank their sperm or eggs, in a facility that kept them frozen for as much as a century, until they were needed. One could imagine a time when many couples produced two generations of children, with their higher fertility rates offsetting those couples who produced no children, taking advantage of nano-enabled medical treatments that prolong both the life span and the portion of it during which women are fertile. Social science research is needed to help us understand whether any technical means can reverse the declining birth rate in the absence of major sociocultural shifts.

Culture and Personality

Survey2000 respondents debated whether the "culture wars" would be settled on the basis of shared ideals and common beliefs, or whether technologies such as the Internet will replace the broadcast media with a babble of narrowcast ideologies and aesthetics. They worried about whether the arts and education would thrive, stimulated by wealth and communication technologies, or whether these fields would fall into an anti-intellectual Philistinism in which corporate greed and popular stupidity combine to stifle creativity. Convergence is very relevant here, partly because it can contribute to rapid progress in technologies of computation, communication, and creativity.

For the past two centuries, the major trend in human artistic culture has been the mass commercialization of music, literature, and related forms of expression. Information technology, however, could reverse this trend. Some would argue that art, music, and literature should always be personal, immediate, and performed live by the original creators themselves. They would return us to the days of wandering minstrels and court jesters. A more practical critique of commercialism notes that thousands of enjoyable novels have lost their copyright protection and are now available free over the web. The same will happen to movies and music recordings over the coming decades.[15]

The idea that government should regulate intellectual property through copyrights and patents is a relatively recent phenomenon in human history, and the precise details of which intellectual property is protected and for how long vary across nations and across history. Two standard sociological justifications for patents or copyrights are cited: They reward creators for their labor, and they encourage greater creativity. Both are empirical claims that can be tested scientifically and could prove to be false in some realms.[16]

Consider music.[17] Star performers existed before the twentieth century, such as Franz Liszt and Niccolo Paganini, but emergence of the mass media has produced a celebrity system promoting a few stars whose music is not necessarily the best or most diverse. Copyrights provide protection for distribution companies and for a few celebrities, thereby helping to support the industry as currently defined, but it may actually harm the majority of performers. This is comparable to Anatole France's famous irony: "The law, in its majestic equality, forbids the rich as well as the poor to sleep under bridges." In theory, copyrights cover the creations of celebrities and obscure performers equally, but only major distribution companies have the resources to defend their property rights in court. In a sense, this is quite fair, because no one wants to steal unpopular music. Nevertheless, by supporting the property rights of celebrities, the copyright concept strengthens them as a class in contrast to more obscure musicians.

Internet music file sharing has become a significant factor in the social lives of many children and young adults, who download bootleg music tracks for their own use and to give as gifts to friends. A 2003 survey by the Pew Internet and American Life Project found that 29 percent of Internet users download music without paying for it, including 52 percent of those aged 18 through 29. Of the users who do so, 67 percent say they do not care about the copyright status of the music.[18]

Clearly, a significant fraction of Americans violate copyright laws, and criminologists would hypothesize they thereby learn contempt for laws in general. Thus, on the level of families, ending copyright could be both morally and economically advantageous. On a much higher level, however, the culture-exporting nations (notably the United States) could stand to lose, although we cannot really predict the net balance of costs and benefits without conducting more research in this area. We do not presently have good cross-national data on file sharing or a well-developed theoretical framework to guide research on whether copyright protection supports cultural imperialism or whether it enhances the positions of diverse cultures in the global marketplace.

Cultural trends will significantly influence the direction nanoconvergence takes, largely because the condition of educational institutions will determine whether competent personnel will be available to develop and implement the technology, and because science education may be reformulated around principles provided and illustrated by nanoscience. Worldwide, great concern has arisen that schools are not giving many children a solid introduction to mathematics and the sciences; at the same time, we have reason to be optimistic that effective methods of teaching are now at hand, based on solid cognitive science research.[19]

Coupled with educational reform, nanoconvergence could transform culture by providing a unified understanding of nature, based on material unity of the chemical, physical, and biological worlds at the nanoscale. This transformation would accomplish what Edward O. Wilson called *consilience* and what Mihail C. Roco and I have called *convergence*: integration of the sciences and other rigorous fields of knowledge, based on shared concepts, language, and research tools.[20] Unification of the sciences would become the basis for unification of other aspects of culture, thereby providing a firm basis of knowledge to both specialists and the general public alike.

Although individuals within any culture inevitably differ, it is also true that the distribution of personality types varies across cultures, and some aspects of personality are an expression of culture.[21] For more than a century, social scientists have argued that technological and economic developments are indirectly but powerfully eroding the traditional basis of human personality by creating a progressively more complex and variable social environment that forces young people to mature under highly inconsistent and individualistic influences.[22] Respondents to *Survey2000* wondered whether people will continue to become increasingly sensate, hedonistic, and self-centered. But will the pendulum actually swing back toward moral conservatism or social responsibility? Or will people become ever more alienated, suffering anomie, neurosis, and even psychosis? Or might a combination of supportive social relationships and improved psychiatric treatments successfully combat the stressful factors pushing toward collective and individual disintegration?

Nanoconvergence may alter psychiatry directly by providing a better understanding of the brain and improved medications, but its indirect influence on mental health—in particular, through its support for economic growth—may actually be greater. Since the mid-nineteenth century, we have known that mental disorders are strongly associated with poverty.[23] As a powerful amplifier of the value of other technologies, nanotechnology can enhance individuals' abilities to attain a diversity of personal goals.

Conversely, if people become more alienated over the coming decades, they will be less trusting and may capriciously oppose new technological developments because of their general suspiciousness of societal institutions rather than a careful analysis of what the technology will actually do. It is unclear whether the value of nanoconvergence will be reduced by a vicious circle of psychopathology, in which bad policy decisions beget more psychopathology, or enhanced by a virtuous circle of nano-enabled economic growth, in which the options for policy makers and ordinary citizens alike improve.

Societal Institutions

The close of the twentieth century gave global dominance to the economic system that its opponents called "capitalism" and its proponents called "free markets" or "private enterprise." Whether this system is currently good or bad, robust or fragile, is a matter of debate, however. Many *Survey2000* respondents suggested that this triumphant economic system will flourish and improve over the coming century, bringing a rising standard of living all over the world and a stable economy, compared with the booms and recessions that characterized previous centuries.

A very different view holds that large conglomerate corporations will control the lives of their workers, owning their homes and the stores where they must buy things. The living conditions and wages will be very poor worldwide, as automation replaces workers and eliminates many of their jobs. Because they have to compete with both these unemployed people and machines that have no bargaining rights, human workers will be forced to accept increasingly bad working conditions.

Much economic inequality reflects differences across nations rather than among individuals within nations, so real improvement of economic well-being must be global in scope.[24] Increased investment in current productive technologies could enable a larger fraction of the world's population to enjoy prosperity, but it is doubtful whether such universal prosperity could be sustained without ruining the environment and exhausting natural resources. Thus, whatever economic system prevails, it must be fueled by technological progress, which will undoubtedly require vastly greater control of nanoscale processes and materials.[25] For example, cheap, clean, renewable sources of energy will be necessary, such as nano-enabled solar power production, vehicles with nanoscale-engineered energy storage media, and vastly increased efficiency of energy use permitted by a wide range of nanotechnologies. Already, many nations have begun working to realize this hope.[26]

Survey2000 respondents disagreed greatly about the future of government, with some maintaining that it will become more benevolent, and others fearing that it will fulfill the nightmares of *Brave New World* and *1984*.[27] Perhaps the criminal justice system will grow ever more civilized, but if the public feels it is drowning in a massive crime wave, the political response could be so mindlessly punitive that it defeats its own purpose of achieving security with justice. Some respondents worried that a future big government will comprise a coalition of exploiters, big and small, who prey upon productive citizens. Others dreamed of a humane welfare state or a workers' paradise. Optimistic moderates voiced hope that democratization and devolution will bring political renewal that can solve the problems of repressive and unresponsive government.

Government establishes the context in which nanoconvergence will either develop or be stifled through its implementation of regulations, its investment in fundamental scientific research, and its support for projects to demonstrate and evaluate particular new applications. In return, nanotechnology can provide new tools by which government can serve the public, especially in areas such as defense and communications between citizens and government, where it amplifies the value of existing technologies. Governments must decide how much they will intrude in nanotechnology-related issues involving other institutions of society, and how intensive a role they will play in evaluating the societal impacts of nanotechnology.

International relations are subject to many stresses—for example, trends related to migration, racism, conflict, warfare, and peace. Some nations have fragmented in recent years; Czechoslovakia and Yugoslavia, for instance, experienced very different processes of division. Conversely, other nations have moved closer together, most notably the member states of the European Union. The fundamental question—how varied the future will be—is partly a matter of the extent to which nations and regional blocs continue to exist. Under a single world government, people might still be divided by non-geographic factors such as religion and ethnicity. If the world is not unified politically, however, the scope of variation and competition could be far greater. Thus it is important to know whether nations are going out of style or whether they are a permanent feature of human life.

Survey2000 respondents disagreed fundamentally about such issues as whether the nations will consolidate to form a single world society or cluster into a small number of competing blocs.[28] Some believed that political independence movements and the failure of governments could split many nations into even smaller units. Will the United States and allied advanced industrial societies be able to maintain their dominance? Or will they fall, only to be replaced by another center of power, such as China, or by world chaos? Years ago, some scholars predicted Japanese dominance, but that has not happened.[29] Obviously, such crucial questions cannot be answered with complete confidence.

Science, Health, and Environment

Many *Survey2000* respondents said that scientific research will continue to achieve significant discoveries, possibly even at an accelerating rate, whereas others suggested that we are nearing the end of the Age of Discovery. Will science soon stall, either because it has exhausted the possibilities for discovery or because society has lost interest in fundamental research? Or will science-based technology improve every individual's spiritual, physical, emotional, and psychological well-being?

Nanotechnology offers to all of the sciences new research instruments, methodologies, and paradigms. Perhaps the greatest potential for radical transformation comes not from any one science, but rather from the integration of all sciences. At the nanoscale, science and technology become one, ensuring that practical applications of new laboratory discoveries can enter the market very quickly. Previously separate disciplines will blend together to form a unified scientific technology that completely controls the structure and properties of manufactured objects, from the atomic level, through the nanoscale level at which large molecules exist, up to the scale of an entire machine, biological–mechanical hybrids, and globe-spanning information systems.

The natural environment is an extremely complex system, and *Survey2000* respondents held a very diverse range of opinions about how it might change over the coming century. Many respondents anticipated that all fossil fuels will run out, causing drastic changes in the distribution of food and manufactured goods, leading to dramatic increases in energy prices, and bringing a halt to industrial development. Others hoped that renewable energy sources will reduce pollution, appropriate technology will humanize industry, and new farming practices will allow more food to be produced. Over the coming century, the conservation movement may grow in strength, until environmentally conscious lifestyles become the norm. At the same time, there is reason to fear that the Earth will suffer massive extinction of species, terrible pollution, global warming, and exhaustion of critical natural resources. These are issues of great importance, public concern, and scientific uncertainty.[30]

Nanoconvergence could potentially help solve these problems by facilitating the development of methods to prevent or remediate conventional forms of pollution, by meeting human goals through novel use of the most abundant and renewable resources, and by improving the efficiency of all technologies, including reducing waste of every kind. For example, nanotechnology may have an important, beneficial role to play in precision agriculture, in which it might potentially improve the quality and quantity of food production while simultaneously reducing pollution. Because some applications of nanoscience may threaten to pollute or to exhaust scarce resources, however, we must be alert for these problems and either avoid these applications or find ways to overcome their environmental disadvantages.

Will human health improve over the coming decades, or will problems of the elderly or of resurgent diseases grow beyond our control? Are the greatest gains to be attained simply through improving people's lifestyle choices, or must we implement aggressive treatments based on genetic engineering and other radical technologies? Both the apparently slowing rate of development of effective new treatments and suggestions from demographers that we are reaching the point of diminishing returns in increasing the life span give cause to doubt that human health and longevity will improve significantly without a radical shift in approach.[31]

Human health can benefit from general improvements in nutrition and public health, made possible by a host of specific nanotechnologies applied to production industries and civil engineering, and from innovations such as nano-enabled microscale sensors to warn of food contamination, disease agents, and harmful pollutants. More controversial techniques will seek to combine nanotechnology with genetic engineering or to introduce nanoscale materials and devices into the human body as means of medical treatment. While we must be careful to avoid increasing risks to human health by blindly adopting all innovations, policy decisions must be made in awareness of the fact that humans are already at great risk of ill health or death. Failing to explore radically new approaches could be the most inhumane choice of all.

HOW WILL THE WORLD BE GOVERNED?

On July 5, 1945, Vannevar Bush presented his influential report, *Science: The Endless Frontier*, to President Harry Truman, thereby setting the stage for funding of scientific research after World War II. Bush is reputed to have been skeptical about the objectivity of the social sciences, but he wrote, "It would be folly to set up a program under which research in the natural sciences and medicine was expanded at the cost of the social sciences, humanities, and other studies so essential to national well-being."[32] Upon reflection, his enthusiasm seems rather tepid, and he does not call for any increase in social science funding.

When the National Science Foundation was established on the basis of Bush's report, the social sciences were not included initially, but they began to receive NSF funding in the late 1950s. By 1980, funding for the social and behavioral sciences had crept up to only 6.0 percent of the NSF budget, but the doctrinaire opposition of the Reagan administration slashed this pittance back to 3.3 percent by 1989.[33] In 2005, the social and behavioral sciences claimed 4.1 percent of the NSF research budget,[34] and NSF accounted for more than 60 percent of basic federal academic research funding in these fields.

Fifty years after Bush submitted his famous report, approximately 350 scientists and engineers met at Research Triangle Park, North Carolina, under the auspices of the science society, Sigma Xi, in a conference called *Vannevar Bush II*. The 1945 report, which came on the heels of the experience of World War II, provided a rationale for U.S. government support of science during the Cold War against the Soviet Union. But the Soviet Union dissolved in 1991 after losing its Eastern European satellites, so a new rationale was needed to justify the U.S. government's role in the twenty-first century. In different ways, the contributors explained that science needed to serve civilian society, just as it had earlier served national defense. Many proclaimed that the social and behavioral sciences would be crucial to success for all scientific fields, because only through scientific research on the dangers and benefits of new technologies could effective policies and investments be decided.

My own contribution was a survey of initiatives that leaders in the social and behavioral sciences had proposed to meet these challenges in seven areas: (1) democratization, (2) human capital, (3) administrative science, (4) cognitive science, (5) high-performance computing and digital libraries, (6) human dimensions of environmental change, and (7) human genetic diversity.[35] This is not applied research at the cost of basic research, but rather fundamental scientific research in strategic areas of special value to the country that offered great opportunities for scientific discovery. Of these seven areas, research in only one has actually been promoted aggressively over the following decade—high-performance computing and digital libraries. Perhaps not surprisingly, this field is the least closely connected to the social and behavioral sciences. In the early 1990s, by contrast, social scientists gave a high priority to research on democratization and the spread of free market economies.

To my surprise (and horror), this research agenda has not been followed. Indeed, the rich societies of the world have failed to invest in the studies that would provide the knowledge needed to secure their future. Politicians advocate their favorite policies in an intellectual vacuum, and the public is treated to mass-media debates from which the social and behavioral sciences have been excluded. I must call this what it is: stupidity.

In the early 1990s, I encountered otherwise thoughtful people who believed that the fundamental questions of politics and economics had been answered. The fall of the Soviet Union, so they thought, proved that the American way is best. They felt sure the rest of the world would follow U.S. leadership into a glorious New World Order of peace and prosperity. Some among them, I sensed, resented social science, as if it were a mild case of treason. To ask scientific questions about democratic institutions and free markets was to question their universal validity. I wonder how many people still feel this way, after years of disappointment, disillusionment, and the emergence of new threats.

The disintegration of the Soviet bloc presented social scientists with unprecedented challenges and opportunities, but the scope of democratization research is actually far broader, extending across the entire world, including the United States, and focusing on the transformation of economic systems as well as political ones. Today, the issue of democratization is an especially salient question in the Middle East and neighboring Islamic countries. At the same time, it should not be forgotten that China still lacks a political system in harmony with its somewhat free economy, and major nations as diverse as Russia and Pakistan have retreated some distance away from democracy. The political futures of many nations, both large and small, remain very much in doubt.

Most Americans continue to support their traditional social, political, legal, and economic institutions, but many harbor grave doubts about both their leaders and their policies. Major domestic social issues being addressed with great difficulty include immigration, health care, crime control, social services, and taxation. Despite decades of progressive legislation, racial and ethnic inequalities persist. In some areas, notably poor inner-city neighborhoods, deprivation seems to have increased. By some measures, the poor are, indeed, getting poorer, and all citizens are concerned about the sluggishness of economic growth. In addition, issues of gender and reproduction continue to divide Americans. Thus a crucial part of democratization research is to examine those factors that will enhance, extend, and sustain democracy in advanced societies such as the United States.

Knowledge in many key scientific areas is so fragmentary at present that research projects promise to achieve great and rapid scientific progress, thereby invigorating entire disciplines. For example, we do not know how tightly connected political liberalization is to free markets, a fact that makes research in politically rigid but economically booming China especially important. The remarkable patience of the people in many post-Soviet states reminds us that we do not really know whether economic collapse and social chaos inevitably lead to political dictatorship, as traditional theories claim.

Social and behavioral scientists supported by NSF have held a number of special workshops (beginning with one about religion I organized in December 1993) to explore the needs and opportunities for research on democratization and the transition to market-oriented economies. Each meeting surveyed the existing body of knowledge in its area, considered the degree of maturity of research methodologies and formal explanatory theories, and identified the most promising specific research topics. Given the more than a decade of neglect that followed the workshops, the issues raised at these meetings are even more pressing today. Brief summaries of reports from six of these workshops follow.

Global Research on the Transformation and Consolidation of Democracies. In this workshop, social scientists from a variety of disciplines developed an over-arching analysis of the needs and opportunities for research on democratization and market transition.[36] Five general research areas deserve special emphasis:

- A host of research questions concern the interplay of economic and political factors around democratization, including the central issue of whether a market economy and economic growth are required for the establishment of democratic institutions.

- The development of legal institutions deserves investigation, because creation and maintenance of democratic government require the rule of law and the complex nexus of norms and roles associated with it.

- Rapid societal change takes place today in a world context, so we must develop a scientific understanding of the global system of both developed and developing nations.

- Given the great economic and cultural diversity of the nations undergoing political change, much new research is needed on the alternative pathways to democracy.

- Systematic study of individual societies can illuminate the pace and the conditions under which transitions to democracy are likely to occur and be sustained.

Religion, Democratization, and Market Transition. Religion is a major force for social and moral change, especially where formal political institutions are undeveloped or compromised.[37] Yet scientific knowledge about the relationships linking religion to democracy and to free markets is highly inadequate. Five topics deserve the highest priority:

- The way in which existing churches and new religious movements promote the development of democratic institutions and free markets in formerly totalitarian societies

- The role religion plays in mature democracies such as the United States as they attempt to sustain themselves and over time to become more democratic

- The ways that religion can become tragically implicated in often bloody conflicts between ethnic groups and nationalist movements

- The religious aspects of immigration and the massive movements of people currently in progress around the world

- The dynamic interplay of social, economic, cultural, and political forces around religion in communities undergoing radical change

Legitimacy, Compliance, and the Roots of Justice. Relatively little scientific research has been completed on the factors that convince people to accept the legitimacy of a political system, of its political institutions and processes, and even of particular government actions, especially outside the United States.[38] Research should address three general types of questions:

- The factors that give popular legitimacy to some governments and governmental actions, while denying it to others

- The sources of breakdown in legitimacy that readily threaten new and poorly rooted democracies or that may even destabilize long-established democratic systems

- The consequences of legitimacy, which may include compliance with official decisions and increased stability of the regime

Geographic Approaches to Democratization. For social geographers, processes of democratization play out most notably in terms of three dimensions: territoriality, spatial structures or flows, and human–environment relations.[39] The amicable division of Czechoslovakia and the relatively trouble-free movement to independence of many parts of the former Soviet Union, for example, took place along preexisting territorial subdivisions. By contrast, the bloodshed in Yugoslavia that occurred during its disintegration and the ongoing war in Iraq remind us of the violence that often attends territorial disputes. The nation-state is a territorial concept, and in many regions of the world ethno-territorial movements challenge existing state structures. On a continental scale, new regional entities are emerging, such as the European Union, and the development of transnational corporations and religious organizations adds to existing complexities. Increased flows of money, technology, and people are eroding old structures and creating new ones. Regionally uneven economic development is a major factor stimulating the migration of people both within and across national boundaries. Such migration and the ensuing demographic changes have profound implications for the development and maintenance of democratic institutions, with the massive refugee problems in several parts of the world now placing extreme stress on already-fragile democratic institutions. Democratization and the development of market-oriented economies have profound implications for management of the environment and natural resources, even before we ask how new technologies may enter into complex world dynamics.

Science in the Cold War and Its Aftermath. Social scientific and historical studies of the ways science was organized and supported during the Cold War can help guide us in reshaping science for the new challenges that lie ahead.[40] From the beginning of World War II, science was mobilized to serve national security, and war-oriented structures have had more than half a century to send their roots deep into U.S. institutions and culture. Among the crucial demands of national security is secrecy; by contrast, the ideal in science is openness of knowledge and free exchange of information. In the past, the high priority given to scientific progress in certain areas vastly increased the number of scientifically trained people and gave the U.S. university system the scope and many of the features that it currently possesses. To some extent, the balance of power and division of labor among various parts of government changed—the presidency versus Congress, the federal government versus state and local governments, defense-related agencies versus administrative units having other functions. Industry adapted to the large defense appropriations and to patterns of international trade shaped by the superpower conflict.

In a very real sense, the Cold War may have created vital partnerships linking universities, industry, and government that supported fundamental scientific research and then exploited the results for the benefit of society. In the twenty-first century, however, the nature of the conflict is likely to be dramatically different from what we faced throughout much of the twentieth century. For this reason, we need careful empirical studies of the current organization and recent history of science so that we can know which parts of the legacy of the Cold War need to be discarded and which preserved.

Science, Technology, and Democracy. Science and technology can be engines of economic development, but their evolution poses a challenging question: What is the most effective balance of government investment versus market mechanisms in both supporting and steering scientific and technological development?[41] To the extent that democratic institutions are made possible and strengthened by prosperity, an important link exists between science and democracy. Likewise, open debate among researchers over scientific questions and the organization of scientific professions can serve as models of democratic governance. In addition, because technical expertise has become essential to public policy making, it is vital to find the right mechanisms through which it might play this role, whether in partnership with elected representatives of the citizenry or through direct public participation. Across the social and behavioral sciences, theoretical models and research methodologies can help develop our understanding of the ways in which science influences, and is influenced by, our democratic institutions and democracy and free economies throughout the world.

A NEW SCIENCE OF SERVICES

In 1994, NSF's Division of Social, Behavioral and Economic Research convened a set of six parallel Human Capital workshops, each of which focused on a question of national concern that was also a very lively area of scientific work, with well-developed research methodologies and formal theoretical models, where aggressive investment in research could pay great dividends to knowledge.[42] The six areas covered by the workshops were the workplace, education, families, neighborhoods, disadvantage, and poverty.

The term "human capital" in this context referred to those acquired non-material resources of a person that allow him or her to participate more effectively in the socioeconomic system. The example that springs to mind most quickly is the valuable skills and knowledge a person may gain through education, but many other kinds of human capital also exist. In the field of sociology, the concept of human capital refers to all those acquirable resources (other than financial capital) that affect the individual's capacity for status attainment. Among these are *social capital,* or the value of the individual's social network plus ascribed social statuses such as race or gender roles, and *cultural capital,* which includes norms, values, and beliefs that are conducive to success. Some research in this area has been supported over the past dozen years, but research now needs to approach the problem of world prosperity from a different direction.

At the third Converging Technologies conference, James Spohrer (Figure 7–1) of IBM urged the development of a new science of service industries. A conventional conception of a modern economy divides it into three sectors: primary (agriculture, fishing, and mining), secondary (manufacturing), and tertiary (services). Roughly speaking, primitive humans participated in only the primary sector (hunting and gathering); more than two centuries ago, the Industrial Revolution initiated a period in which manufacturing dominated in terms of money and workers; and the balance shifted to services in the twentieth century. By 2005, 78.7 percent of the U.S. economy was said to be in services, versus only 20.4 percent in industry and the remainder in agriculture.[43]

I am somewhat skeptical of the current utility of this division into sectors and, therefore, about any estimates of the relative balance between manufacturing and services. For example, as an author of books, articles, and software, I wonder whether writing is a service or an act of manufacturing. Some software designers and programmers are employed on a "work on hire"

Figure 7–1   James Spohrer, Director of Almaden Services Research at IBM's Almaden Research Center. He helped launch and has consistently supported the converging technologies effort, and is a leading innovator in the development of a new science of service industries. (Image reproduced by permission of IBM Research, Almaden Research Center. Unauthorized use not permitted.)

basis by corporations that use the programs they create; this sounds like a service. But others write software that will be mass produced and sold to thousands of individual users; this sounds like manufacturing. Even though services are clearly important in the modern economy, they nevertheless present a challenge to economic progress, because it has proven much more difficult to increase productivity in services than in manufacturing. Partly this discrepancy stems from the efficiency of mass production; services are much less efficient because they are often tailored or delivered separately for each specific client.

One of my Harvard professors, Daniel Bell, argued in 1973 that the rise of services would open a new era he called the *postindustrial society*.[44] In this brave new world, the professional and technical class would be preeminent, and theoretical knowledge would be central to innovation and policy formulation. Efforts to shape the future would harness technology, be guided by technology assessment, and lead to the creation of new "intellectual technology" for making decisions. To guide society, scientists would measure key social indicators, mobilizing human resources in four areas:

- The social costs and net benefits of innovations
- The nature and magnitude of social ills, such as crime and family disruption
- Performance in meeting social needs, such as housing and education
- Opportunities for socioeconomic mobility

I recall a really stunning debate between Bell and Harrison White, the physicist-sociologist mentioned in Chapter 6. This contest took place in the third-floor lounge of William James Hall, Harvard's home for the social and behavioral sciences. In an intimate afternoon tea hour, Bell expounded on the postindustrial society. Then White began debating his methods, claiming they were not really scientific. Bell referred to his concepts as lenses or prisms through which to view reality, but White suggested they were little more than personal opinions. To be sure, Bell's style was similar to that of Talcott Parsons, although his prose was vastly more lucid, and he wrote commentaries on contemporary society rather than carrying out empirical studies to test hypotheses. Although Bell's ideas were very interesting, I thought White had done better in their debate. The final judgment would be rendered by the test of time.

Recalling Bell's and White's arguments decades later, I sense that Bell was both right and wrong. Indeed, information technologies had become far more important with each passing year, and services continued to expand. The increasing complexity of society and of social problems, including globalization and the incomplete process of democratization, implies that we need rational solutions based on well-developed technical knowledge. However, the professional and technical class is not preeminent. Rather, the most powerful people are entrepreneurs, executives, and politicians. They use the expertise of scientists and engineers when it pleases them to do so, and then they blithely cast the experts aside. I tend to think Bell was correct about what *should* happen, but not about what *did* occur.

Spohrer's vision of a future services science could potentially turn this trend in a new direction, by conceptualizing the entire economy in terms of services guided by information technology. He notes that work depends on capabilities, which are practical abilities and plans to achieve a goal to create value. According to Spohrer, "For humans, capabilities come in four basic types: technological (tools), social (relationships, organizations), cognitive (skills, attitudes, ideas), and environmental (nature, useful spaces, culture)."[45] For any task, there are several ways to organize work practices. For example, different degrees of emphasis can be placed on the tool system (harnessing nature following technosocial models with stochastic parts) versus the human system (organizing people following socioeconomic models with intentional agents).

In collaboration with IBM's Doug Riecken, Spohrer edited the special July 2006 issue of the journal *Communications of the Association for Computing Machinery*, which is dedicated to services science. Several contributors stressed that services organizations should emphasize rapid innovation.[46] When the focus shifts to information technology, the distinction between manufacturing and services vanishes. In their article, Roland Rust and Carol Miu of the University of Maryland argued that competitiveness in the global economy requires all businesses to become service businesses.[47] Perhaps left unsaid in these enthusiastic essays is the need for a resurrected social science, firmly based in cognitive and information sciences, to form the fundamental core of services science.

In his original *Converging Technologies* essay on services science, Spohrer suggested that much fabrication of goods in future will be done locally, using simple ingredients with methods such as stereolithography (i.e., "three-dimensional printing"). This process entails a collection of methods for building up or cutting down raw materials under automatic computer control. Today it is chiefly used for rapid prototyping, because mass production still makes most products more cheaply. Spohrer suggests that the vastly increased understanding of materials, gained from nanoscience, could reverse this economic disadvantage. Under these circumstances, a small workshop in a small town could produce thousands of different items cheaply, using relatively small equipment guided by a computer containing the designs.

In the same NBIC book, Bruce Tonn (Figure 7–2a) suggests this break-through could produce economies that are much more decentralized and sustainable, are based on small-scale social organizations, and thus are more humane.[48] Tonn is a perfect example of a social scientist who has achieved convergence in his own career. He has published extensively on the environment, the economy, and the future of social trends that are influenced by innovation. Notably, he has argued that humanity may have outgrown the principle that governments must be based on the control of particular realms of land. Instead, Tonn suggests, modern communication technologies and the diversity of modern cosmopolitan societies may render nonspatial government both feasible and desirable. Governmental authority could be devolved not only back to smaller geographical units, but also to widely dispersed communities that are based on shared interests rather than territorial boundaries.[49] These possibilities raise age-old questions about how disputes will be handled when competing groups claim jurisdiction over the same population or area of interest, and whether contemporary politics is making a graceful transition to new forms or merely defending old institutions of power against change.

Sociologist and ethicist James Hughes (Figure 7–2b) argues that biotechnology and the NBIC convergence are creating a third dimension of debate that could dominate the politics of the twenty-first century.[50] In the previous

Figure 7–2 Bruce Tonn (a) and James Hughes (b). These two rigorous but visionary social scientists have been exploring the human implications of converging technologies, including the potential cultural and political conflicts that may emerge, and the potential need for new societal institutions to achieve the most beneficial changes.

century, debates were largely two-dimensional: The economic dimension dealt with issues of inequalities, labor, welfare, and taxation, whereas the cultural dimension extended across questions related to race, gender, nationalism, and civil liberties. The new third dimension pits technoconservatives and technoluddites against technolibertarians and transhumanists. One implication of this analysis is that the coalition of people supporting rapid technological convergence may bring together individuals with highly diverse views on the issues that define the two other dimensions. Hughes believes that society may become polarized along the technopolitical dimension through disputes over such concrete issues as antiaging research, gene therapies, prenatal care outside the womb, intellectual enhancement of animals, regulation of the risks of NBIC applications such as nanomaterials, parental rights to use genetic testing to select "better" children, and proliferation of "cyborg" implantable information technologies.

CONCLUSION

Critics of convergence, advocates of convergence, social scientists, and ordinary people have put forth wildly disparate images of the human future. In itself, this splintered picture of things to come suggests that the future will be infinitely complex. One would guess that the chaotic result of all these practically unpredictable factors would be the emergence of subcultures, whether based in local communities or connected to shared interests and interactions that are independent of geography. If the arts become decommercialized, then they will be free again to express the values of local communities and topical subcultures.

If Spohrer and Tonn are right, local NBIC workshop manufacture might strengthen local economies at the expense of large corporations and regional economies. Economies of scale demand concentration of manufacturing and economic power in a few hands, but efficiency is not the only source of value. More important is the production of goods and provision of services that actually meet people's individual needs. Perhaps this goal can be achieved better through local and diversified manufacture, which in turn is made more efficient by nanoconvergence. Of course, efficiency is not the only source of concentration; other key areas include monopolization of patents and other intellectual property, mass advertising and brand loyalty, and implicit government favoritism in regulations and contracts. Thus devolution of industry to local workshops is not a forgone conclusion and would require substantial socioeconomic changes.

The emergence of a peaceful, progressive world society depends—at the very least—on a technology that can offer abundance to a majority of people in all societies. Even in the short term, nanotechnology can be an effective multiplier of the effectiveness of many other wealth-producing technologies. In the long term, manufacturing through control of matter at the nanoscale might potentially turn out to be one of the greatest treasures of all mankind. In a great variety of ways, nanotechnology may increase the effectiveness of military forces; the question then becomes whether those forces will be fighting for or against civilization, freedom, and justice. In convergence with other sciences, nanoscience could become the intellectual basis for a shared, global culture, thereby helping nations to cooperate on development projects.

Nanoscience and nanotechnology could have significant and wide-ranging benefits across major areas of human life where powerful trends are already bringing change: family, culture, personality, economy, government, nations, science, environment, and health. My comments in this chapter have been based on theory more than empirical evidence, however. Clearly, it will be essential to work out appropriate research methods for examining nano-assisted change in all of these realms, and to launch well-designed research projects at the earliest feasible time.

Only by starting soon can we realize the tremendous scientific advantage of observing revolutionary technological change at an early stage in its development. Only by examining the societal implications of nanotechnology in many varied settings can we understand the full potential of this multifaceted world of science and engineering. And only by creating a new social science, unencumbered by the ideologies and disappointments of the past, and based in the convergence of cognitive and information sciences, will we be able to understand and steer the future course of a transformed world.

REFERENCES

1. Alex Inkeles and David H. Smith, *Becoming Modern: Individual Changes in Six Developing Countries* (Cambridge, MA: Harvard University Press, 1974); Alejandro Portes, "The Factorial Structure of Modernity," *American Journal of Sociology*, 79:15–44, 1973.

2. Richard R. Nelson, "Diffusion of Development: Post–World War II Convergence Among Advanced Industrial Nations," *American Economic Review*, 81(2):271–275, 1991; Bruce Truitt Elmslie, "The Convergence Debate Between David Hume and Josiah Tucker," *Journal of Economic Perspectives*, 9(4):207–216, 1995; Andrew B. Bernard and Charles I. Jones,

"Technology and Convergence," *Economic Journal*, 106:1037–1044, 1996; Samuel P. Huntington, *The Cash of Civilizations and the Remaking of World Order* (New York: Simon and Schuster, 1996).

3. Jeffrey K. Liker, Carol J. Haddad, and Jennifer Karlin, "Perspectives on Technology and Work Organization," *Annual Review of Sociology*, 25:575–596, 1999; Mauro F. Guillen, "Is Globalization Civilizing, Destructive or Feeble?" *Annual Review of Sociology*, 27:235–260, 2001.

4. Malcolm I. Thomis, *The Luddites: Machine-Breaking in Regency England* (New York: Schocken Books, 1970).

5. http://www.archive.org/index.php

6. ETC Group, *The Big Down: Atomtech—Technologies Converging at the Nano-scale* (Ottawa, Canada: ETC Group, 2003, p. 6), online at http://www.etcgroup.org/documents/TheBigDown.pdf

7. ETC Group, *The Big Down: Atomtech—Technologies Converging at the Nano-scale* (Ottawa, Canada: ETC Group, 2003, p. 5), online at http://www.etcgroup.org/documents/TheBigDown.pdf.

8. Toby Shelley, *Nanotechnology: New Promises, New Dangers* (London: Zed Books, 2006).

9. Daniel Chirot, *Social Change in the Twentieth Century* (New York: Harcourt Brace Jovanovich, 1977).

10. William Sims Bainbridge, *Goals in Space* (Albany, NY: SUNY Press, 1991).

11. William Sims Bainbridge, "Religious Ethnography on the World Wide Web," in Jeffrey K. Hadden and Douglas Cowan (eds.), *Religion and the Internet* (Greenwich, CT: JAI Press, 2000, pp. 55–80); "The Spaceflight Revolution Revisited," in Stephen J. Garber (ed.), *Looking Backward, Looking Forward* (Washington, DC: National Aeronautics and Space Administration, 2002, pp. 39–64); "Massive Questionnaires for Personality Capture," *Social Science Computer Review*, 21:267–280, 2003; "The Future of the Internet: Cultural and Individual Conceptions," in Philip N. Howard and Steve Jones (eds.), *Society Online: The Internet in Context* (Thousand Oaks, CA: Sage, 2004, pp. 307–324).

12. https://www.cia.gov/cia/publications/factbook/index.html

13. Kingsley Davis, Mikhail S. Bernstam, and Rita Ricardo-Campbell (eds.), *Below-Replacement Fertility in Industrial Societies: Causes, Consequences, Policies* (New York: Population Council, 1986); Ben J. Wattenberg, *The Birth Dearth* (New York: Ballantine, 1987); cf. Kingsley Davis, "The Theory of Change and Response in Modern Demographic History," *Population Index*, 29:345–366, 1963.

14. U.S. Central Intelligence Agency, *Long-Term Global Demographic Trends: Reshaping the Geopolitical Landscape* (Langley, VA: CIA, OTI IA 2001-045, 2001).

15. Robert Graves, *Watch the North Wind Rise* (New York: Creative Age Press, 1949).

16. Carole A. Ganz-Brown, "Electronic Information Markets: An Idea Whose Time Has Come," *Journal of World Intellectual Property*, 1:465–494, 1998; National Research Council, Computer Science and Telecommunications Board, *The Digital Dilemma: Intellectual Property in the Information Age* (Washington, DC: National Academy Press, 2000).

17. William Sims Bainbridge, "New Technologies for the Social Sciences," in Marc Renaud (ed.), *Social Sciences for a Digital World* (Paris: Organisation for Economic Co-operation and Development, 2000, pp. 111–126).

18. Mary Madden and Amanda Lenhart, "Music Downloading, File-Sharing and Copyright," *Pew Internet and American Life Project*, 2003, online at http://www.pewtrusts.com/pdf/pew_internet_file_sharing_073103.pdf

19. Ina V. S. Mullis, Michael O. Martin, Eugenio J. Gonzalez, Kathleen M. O'Connor, Steven J. Chostowski, Kelvin D. Gregory, Robert A. Garden, and Teresa A. Smith, *Mathematics Benchmarking Report: TIMSS 1999— Eighth Grade* (Chestnut Hill, MA: International Study Center, Lynch School of Education, Boston College, 2001); Michael O. Martin, Ina V. S. Mullis, Eugenio J. Gonzalez, Kathleen M. O'Connor, Steven J. Chostowski, Kelvin D. Gregory, Teresa A. Smith, and Robert A. Garden, *Science Benchmarking Report: TIMSS 1999–Eighth Grade* (Chestnut Hill, MA: International Study Center, Lynch School of Education, Boston College, 2001); John D. Bransford, Ann L. Brown, and Rodney R. Cocking (eds.), *How People Learn: Brain, Mind, Experience, and School* (Washington, DC: National Academy Press, 1999).

20. Edward O. Wilson, *Consilience: The Unity of Knowledge* (New York: Random House, 1998); Mihail C. Roco and William Sims Bainbridge, *Converging Technologies for Improving Human Performance* (Dordrecht, Netherlands: Kluwer, 2003).

21. Marvin K. Opler (ed.), *Culture and Mental Health: Cross-cultural Studies* (New York: Macmillan, 1959); Ari Kiev, *Transcultural Psychiatry* (New York: Free Press, 1972).

22. Ferdinand Tönnies, *Community and Society* (East Lansing, MI: Michigan State University Press, 1887 [1957]); Emile Durkheim, *The Division of Labor in Society* (New York: Free Press, 1893 [1964]); Emile Durkheim,

Suicide (New York: Free Press, 1897 [1951]); Robert E. L. Faris and H. Warren Dunham, *Mental Disorders in Urban Areas* (Chicago: University of Chicago Press, 1939); David Riesman, *The Lonely Crowd: A Study of the Changing American Character* (New Haven, CT: Yale University Press, 1950); Philip Elliot Slater, *The Pursuit of Loneliness: American Culture at the Breaking Point* (Boston: Beacon Press, 1970).

23. Edward Jarvis, *Report on Insanity and Idiocy in Massachusetts* (Boston: White, 1855); cf. August B. Hollingshead and Frederick C. Redlich, *Social Class and Mental Illness* (New York: Wiley, 1958).

24. Glenn Firebaugh, "Empirics of World Income Inequality," *American Journal of Sociology,* 104:1597–1630, 1999; "The Trend in Between-Nation Income Inequality," *Annual Review of Sociology,* 26:323–339, 2000; Harland Cleveland, "The Global Century," *Futures,* 31:887–895, 1991.

25. Mihail C. Roco, R. S. Williams, and P. Alivisatos, *Nanotechnology Research Directions* (Dordrecht, Netherlands: Kluwer, 2000).

26. Richard W. Siegel, Evelyn Hu, and M. C. Roco, *Nanostructure Science and Technology* (Dordrecht, Netherlands: Kluwer, 1999).

27. Aldous Huxley, *Brave New World* (Garden City, NY: Doubleday, Doran, 1932); George Orwell, *Nineteen Eighty-Four, a Novel* (New York: Harcourt, Brace, 1949).

28. Samuel P. Huntington, *The Clash of Civilizations and the Remaking of World Order* (New York: Simon and Schuster, 1996).

29. Ezra F. Vogel, *Japan as Number One: Lessons for America* (Cambridge, MA: Harvard University Press, 1979).

30. Robert T. Watson, Marufu C. Zinyowera, Richard C. Moss, and David J. Dokken (eds.), *Climate Change 2001* (Geneva, Switzerland: Intergovernmental Panel on Climate Change c/o World Meteorological Organization, 2001), www.ipcc.ch, www.grida.no/climate/ipcc/regional/index.htm

31. Samuel H. Preston, "American Longevity: Past, Present, and Future" (Syracuse, NY: Center for Policy Research, Syracuse University, 1996); Victor R. Fuchs, "Health Care for the Elderly: How Much? Who Will Pay for It?" *Health Affairs,* 18:11–21, 1999; Frederick W. Hollmann, Tammany J. Mulder, and Jeffrey E. Kallan, "Methodology and Assumptions for the Population Projections of the United States: 1999 to 2100," Population Division Working Paper No. 38 (Washington, DC: U.S. Census Bureau, 2000); cf. Richard H. Steckel and Roderick Floud (eds.), *Health and Welfare During Industrialization* (Chicago: University of Chicago Press, 1997).

32. Vannevar Bush, *Science: The Endless Frontier* (Washington, DC: U.S. Government Printing Office, 1945, p. 18).

33. Otto N. Larsen, *Milestones and Millstones: Social Science at the National Science Foundation, 1945–1991* (New Brunswick, NJ: Transaction, 1992, p. 173).

34. National Science Foundation, FY 2007 Budget Request to Congress, http://www.nsf.gov/about/budget/fy2007/toc.jsp

35. William Sims Bainbridge, "Strategic Research in the Social Sciences," in John F. Ahearne (ed.), *Vannevar Bush II: Science for the 21st Century* (Research Triangle Park, NC: Sigma Xi, 1995, pp. 207–226).

36. James L. Gibson (ed.), *Democratization: A Strategic Plan for Global Research on the Transformation and Consolidation of Democracies*, 1993, online at http://www.nsf.gov/sbe/ses/soc/works4.jsp

37. John McCarthy (ed.), *Religion, Democratization, and Market Transition*, 1993, online at http://www.nsf.gov/sbe/ses/soc/works3.jsp

38. Gregory Caldeira (ed.), *Legitimacy, Compliance, and the Roots of Justice: A Strategic Plan for Global Research* (Arlington, VA: National Science Foundation, 1994).

39. Alexander B. Murphy (ed.), *Geographic Approaches to Democratization: A Report to the National Science Foundation* (Eugene, OR: University of Oregon Press, 1995).

40. David A. Hounshell (ed.), *Science, Technology, and Democracy in the Cold War and After: A Strategic Plan for Research in Science and Technology Studies*, online at http://www.cmu.edu/coldwar/NSFbookl.htm

41. Paul B. Thompson (ed.), *Science, Technology, and Democracy: Research on Issues of Governance and Change* (College Station, TX: Texas A&M University, 1994).

42. Rebecca Blank (ed.), *Investing in Human Resources: A Strategic Plan for the Human Capital Initiative*, 1994, online at http://www.nsf.gov/sbe/ses/soc/works1.jsp

43. https://www.cia.gov/cia/publications/factbook/geos/us.html

44. Daniel Bell, *The Coming of Post-industrial Society* (New York: Basic Books, 1973).

45. James Spohrer, Douglas McDavid, and Paul P. Maglio, "NBIC Convergence and Technology–Business Coevolution: Towards a Services Science to Increase Productive Capacity," in William Sims Bainbridge and Mihail C. Roco (eds.), *Managing Nano-Bio-Info-Cogno Innovations: Converging Technologies in Society* (Berlin: Springer, 2006, p. 230).

46. Henry Chesbrough and James Spohrer, "A Research Manifesto for Services Science," *Communications of the ACM*, 49(7):35–40, 2006; Jerry Sheehan, "Understanding Service Sector Innovation," *Communications of the ACM*, 49(7):43–47, 2006; William B. Rouse and Marietta L. Baba, "Enterprise Transformation," *Communications of the ACM*, 49(7):67–72, 2006.

47. Roland T. Rust and Carol Miu, "What Academic Research Tells Us about Service," *Communications of the ACM*, 49(7):49–54, 2006.

48. Bruce E. Tonn, "Coevolution of Social Science and Emerging Technologies," in William Sims Bainbridge and Mihail C. Roco (eds.), *Managing Nano-Bio-Info-Cogno Innovations: Converging Technologies in Society* (Berlin: Springer, 2006, pp. 309–335).

49. Bruce Tonn and D. Feldman, "Non-spatial Government," *Futures*, 27(1):11–36, 1995.

50. James J. Hughes, "Human Enhancement and the Emergent Technopolitics of the 21st Century," in William Sims Bainbridge and Mihail C. Roco (eds.), *Managing Nano-Bio-Info-Cogno Innovations: Converging Technologies in Society* (Berlin: Springer, 2006, pp. 285–307).

Chapter 8

The Final Frontier

Previous chapters in this book have shown that nanoconvergence can achieve real but moderate improvements in most fields of technology, and can support continued economic growth and the betterment of human welfare, if applied wisely. However, we have not decided whether nano is truly revolutionary or merely the next step in the consolidation of human technical knowledge and capabilities. To see beyond the next decade or two, and to envision truly revolutionary changes in the conditions of human life, we need a perspective that is both lofty and realistic. This chapter accomplishes that goal by examining the preeminent example of technological transcendence— the possibility that nano-enabled space travel will allow our civilization literally to rise above the mundane limits of our home planet.

THE GIANT LEAP

Ever since Neil Armstrong stepped on the surface of the Moon in 1969, there has been debate about exactly what words he spoke on that historic occasion. In 2006, computer analysis of his recorded voice verified that Armstrong had said, "That's one small step for a man, one giant leap for mankind."[1] But was it? Perhaps spaceflight was not a leap, but rather a stumble. In my 1976 book *The Spaceflight Revolution,* I wrote, "Either spaceflight will be proven a successful revolution that opened the heavens to human use and habitation, or it will be proven an unsuccessful revolution that demonstrated in its failure the limits of technological advance."[2]

Narrowly defined in terms of human travel to other worlds, the Space Age lasted from December 1968 through December 1972, from the first flight of Apollo 8 around the Moon to the last flight of Apollo 17 to its surface. For decades, no human being has flown higher than 400 miles above the Earth's surface. When the space shuttle was developed in the 1970s, it was hoped it could reliably deliver humans and equipment to Earth orbit at low cost on a regular schedule. But first the Challenger launch disaster of 1986 and then the Columbia reentry disaster of 2003 proved to the wider public what close observers of the shuttle program already knew: It was the worst kind of failure—a project that succeeded just well enough to continue, but was hugely more costly than expected and too complex ever to be reliable.

In the wake of the second shuttle loss, the United States was forced to come to terms with these grim facts. The decision was made to fly just enough shuttle flights to complete the International Space Station. Emphasis shifted to the development of new launch vehicles, based on the same general principles as those supporting the vehicles used in the Apollo program, but enhanced by whatever improvements in materials and design were made possible by four decades of general technological progress. "Apollo on steroids," some called it. Development of the new vehicles would be costly, so many valuable research projects were cancelled, to the great consternation of many scientists.[3] Some politicians and members of the general public may have been pleased that the new vehicles were designed for a return to the Moon and eventual manned missions to Mars. Critics, including some NASA insiders speaking privately, criticized the vagueness of the new program's research goals and predicted that the new vehicles would be delivered late, over budget, and with the same mixture of marginal performance and unreliable complexity as had plagued the shuttle.

In economic terms, spaceflight is costly and offers a relatively small return on most investments. Early accomplishments were largely the result of a transcendent social movement that was able to exploit international competition to support its goals, without caring very much about economic factors. A resurgence of space exploration could be launched by nanotechnology advances, if they simultaneously reduce the cost and increase the profit. Importantly, it may be necessary to find entirely new uses for outer space and new motivations for undertaking voyages there.[4]

Nanotechnology advances, especially in convergence with the other NBIC fields, may reduce the cost and increase the reliability of space travel, thereby tilting the cost–benefit balance toward the positive. The very first conference on the societal implications of nanotechnology observed:

> The stringent fuel constraints for lifting payloads into earth orbit and beyond, and the desire to send spacecraft away from the sun for extended missions (where solar power would be greatly diminished), compel continued reduction in size, weight, and power consumption of payloads. Nanostructured materials and devices promise solutions to these challenges. Nanostructuring is also critical to the design and manufacture of lightweight, high-

strength, thermally stable materials for aircraft, rockets, space stations, and planetary/solar exploratory platforms.[5]

At the same conference, Samuel Venneri, Chief Technologist of NASA, argued that all of the easy missions had already been carried out, and that the next stages of space exploration would require significant technological advances. Notably, robot space probes will need to operate far from Earth without constant human supervision in hostile environments like the supposed ocean under the ice of Jupiter's moon Europa, or the thick atmospheres of Venus and Saturn's moon Titan. Even before the conference had explicitly recognized the centrality of NBIC technological convergence, Venneri argued in its favor:

> [Our] current challenge is to develop space systems that can accomplish our missions effectively and economically. To do this they will have to be much more capable than today's spacecraft. They will have to have the characteristics of autonomy to "think for themselves"; self-reliance to identify, diagnose, and correct internal problems and failures; self-repair to overcome damage; adaptability to function and explore in new and unknown environments; and extreme efficiency to operate with very limited resources. These are typically characteristics of biological systems, but they will also be the characteristics of future space systems. A key to developing such spacecraft is nanotechnology.[6]

NASA was one of the agencies that created the National Nanotechnology Initiative (NNI), standing in fourth place with a budgetary commitment of $46 million in fiscal year 2002, behind NSF ($199 million), the Department of Defense ($180 million), and the Department of Energy ($91 million), but ahead of NIH ($41 million), the National Institute of Standards and Technology ($38 million), and four other agencies with very small commitments. In 2002, NASA explicitly endorsed the vision of convergence, even using that term in its formal statement of expected benefits:

> NASA's driving considerations for developing advanced technology in pursuit of the agency's goals and missions are to reduce cost, increase safety, and achieve greater performance. These objectives will be enabled through the application of nanotechnology in the development of new generations of safer, lighter, and more efficient vehicles. NASA believes the convergence of nanotechnology, biotechnology, and information [technology] will provide unprecedented benefits and solutions to its myriad

mission challenges. Nanostructured materials and devices promise solutions to many of these challenges. Moreover, the low-gravity, high-vacuum space environment may aid development of nanostructures and nanoscale systems that cannot be created on Earth. Applications include (a) low-power, radiation-tolerant, high-performance computers; (b) nano-instrumentation for micro-spacecraft; (c) wear-resistant nanostructured coatings; (d) enhanced safety through deployment of novel sensors and electronic mentors; and (e) in the long term, technologies leading to the development of intelligent autonomous spacecraft.[7]

Although NASA did not use the term "cognitive technologies" in this description, intelligent autonomous spacecraft will clearly require them (we will examine this topic more deeply later in this chapter). NASA's convergent vision for nanotechnology persisted, as seen in the statement it contributed to the NNI supplement to the President's fiscal year 2006 budget request, in the area of "fundamental nanoscale phenomena and processes":

> NASA: Pursue the strategy of using the biological model of hierarchical cellular organization for the application of nanotechnology and for the management of information as a paradigm for future space systems and explorers. Using real-time feedback loops, conduct research at the intersection of biology and nanotechnology to develop (a) a bio-analytical laboratory for interrogating extraterrestrial samples; (b) high-throughput, quantitative physiological monitoring for astronauts; and (c) diagnostic technologies for spaceship environmental monitoring. For long-term research, the goal is to capitalize on nature's model for the management of information, achieving emergent complex functionality through the self-organization of a large number of primitive information inputs.[8]

In 2006, the National Nanotechnology Coordination Office published a report, titled *Nanotechnology in Space Exploration*, that summarized the actual potential of nanotechnology, often in convergence with other fields, to improve the efficiency, effectiveness, and reliability of space technologies. The report cited many areas where nanoconvergence could improve the effectiveness of space technology:[9]

Near to Mid-Term (5–10 Years)

- New lightweight materials with superior strength compared to conventional ones, useful for space and aeronautic systems

- New adhesives and thermal protection materials
- Materials with enhanced radiation protection, including the capability of measuring radiation (embedded radiation dosimeters)
- New electronic components and high-density memory devices
- Ultra-sensitive and selective sensor devices
- New materials for vehicle and human health management

Long Term (Beyond 10 Years)

- New fault-tolerant computing and communication technologies
- Microcraft for autonomous exploration
- Multifunctional materials offering thermal, radiation, and impact resistance
- New approaches for energy generation, storage, and distribution
- Hierarchical systems exploiting developments in biotechnology, information technology, and nanotechnology
- Environmentally friendly methods for biomimetic synthesis

Notice that the last two items in the list of areas ripe for long-term development explicitly depend on convergence.

Unfortunately, the new vision for space exploration of the U.S. government deemphasizes scientific research and technological advances, and instead assigns the highest priority to operational missions using either existing technology or the upgraded version of 1960s Apollo technology that is intended for the next two decades. This preference suits the aerospace companies just fine, because they will be able to build multiple identical booster-stage rockets that are destroyed when they are used. Many observers of NASA think it should be a research and development agency that focuses on developing new technologies for the future, rather than being primarily tasked with conducting operational missions. Unfortunately, nanotechnology budgets in NASA are shrinking, from $46 million in fiscal year 2002 to a requested $25 million in 2007.[10]

THE REALITIES OF INTERPLANETARY TRAVEL

Spaceflight is extremely difficult, at the present level of technology. To achieve low Earth orbit, a spacecraft must reach a speed of around 8 kilometers per second (approximately 5 miles per second). This is 9 times the velocity of bullets fired from an M16 rifle and 30 times the speed of a jet airliner. Rockets suffer from an inherent inefficiency—namely, they must accelerate their fuel as well as their payload. A bullet does not carry its gunpowder with it, nor does an aircraft carry the oxygen from the surrounding atmosphere with which it will burn its fuel. As rocket pioneers of the early twentieth century realized, a feasible but very expensive way to circumvent this inefficiency is to construct rockets in two or three stages, employing a large booster to accelerate a smaller rocket before dropping away.

Among the best propellants are oxygen and hydrogen, which are used in the main engines of the space shuttle. These elements are among the most energetic chemicals, yet are environmentally benign—their reaction simply produces water. However, they must be stored as liquids, meaning a temperature of approximately −253 °C (−423 °F) for hydrogen and −183 °C (−297 °F) for oxygen. Without the polluting solid rocket boosters serving as a first stage, the space shuttle could hardly lift off the launch pad, let alone achieve orbit. It is possible to imagine a future single-stage oxygen–hydrogen launch vehicle, with most of the structure composed of strong and light carbon nanotubes, reaching orbit without a booster. At present, we are nowhere near ready to build such a machine.

In the 1960s, experiments were done with nuclear rockets. This idea was first conceived by German rocket engineers in the early 1940s, who proposed running hydrogen through a hot reactor to achieve thrust. Clearly, the public would need to be persuaded to view nuclear energy in a much more favorable light before it would permit development of spaceships based on nuclear reactors or explosions, and one dramatic accident could be enough to terminate such a program.

A very different approach would be to build a launch vehicle based on hypersonic jet engine propulsion. Although a single-stage-to-orbit vehicle is conceivable, optimization of design details would be extremely difficult, so a two-stage model is easier to contemplate. The larger boost stage would consist of a reusable jet plane designed for high speed, propelled by a supersonic ram jet (i.e., scramjet) that can operate much faster than the complex turbo jets with spinning compressors that currently propel jet airliners. Because ramjets cannot begin working until the vehicle is already moving fast enough to compress the air before combustion, take-off might require a high-speed catapult. The advantage of the scramjet over a rocket is that it would use the oxygen from the atmosphere to burn its fuel, rather than having to carry it aboard. A disadvantage is that this vehicle needs to stay in the atmosphere until it reaches a good fraction of orbital velocity, which poses a number of design challenges, such as variable geometry of the airframe or air intakes, active cooling of key engine surfaces, and careful design of how the orbital second stage is carried and protected from the high-speed travel through the atmosphere.

In 2004, the National Research Council published a highly informative but ultimately melancholy report, *Evaluation of the National Aerospace Initiative*, about cooperation between NASA and the U.S. Department of Defense on launch vehicle development.[11] The report expressed optimism about the feasibility of hypersonic scramjet launch vehicles, but suggested that a decade's aggressive research would be required to determine whether such a vehicle was really possible, and then many further years of work to actually produce one. Presumably, nanotechnology would provide some of the needed materials, and the final design would rely very heavily on information technology; the two other NBIC fields would probably not be relevant. The melancholy quality of the report stems not only from awareness of the huge difficulty presented by the scientific and engineering challenges, but also from the realization that NASA was going in a very different direction in its plans for new launch vehicles, so funding would most likely not be available to realize this dream. It is also worth keeping in mind that NASA failed in at least two prior attempts to produce new reusable launch vehicles that would replace the space shuttle—the National Aero-Space Plane of the 1980s, and the X-33 in the late 1990s.[12]

Once orbit has been achieved, a number of propulsion options are available to take humans, robots, and materials to extraterrestrial destinations. For example, solar energy can be used to drive ion rocket engines, which have low thrust but are extremely efficient. Of all the places in the solar system that humans or their machines might visit, only Mars, Venus, and Saturn's moon Titan have both atmospheres and solid surfaces, and only Mars is a plausible destination for humans in this century. Without an atmosphere, as is the case with the Moon, rocket fuel must be carried for landing as well as for returning. If robots serve as our explorers, the huge investment in fuel and equipment for the return would be unnecessary, of course. Robert Zubrin has argued that fuel could be created on Mars for the return journey, and it would make sense to send robots to complete that crucial mission before human explorers even launched.[13]

Although outer space is chiefly a vacuum, it does contain hazards. Space junk—that is, fragments of artificial satellites or trash accidentally released by the space shuttle—is one of the main dangers in Earth orbit. Meteoroids are rare near the Earth, especially large ones, but they typically move very fast in relation to a spacecraft and thus may cause significant damage. During its flight in September 2006, the shuttle Atlantis was struck by some kind of debris, which punched a hole one-tenth of an inch in diameter in a nonessential heat radiating panel inside one of the open payload doors. The culprit was very unlikely to be jetsam from the September mission itself, because jetsam would have been traveling in the same orbit and thus moving too slowly relative to the spacecraft to cause such a clean hole.[14] Beyond Earth orbit, meteoroids are a real if uncommon threat.

Dense concentrations of meteoroids can be predicted and avoided—most notably, those that cause meteor showers on Earth, like the Perseids and Leonids. On August 12–14, 1993, cosmonauts Vasili Tsibliev and Alexander Serebrov experienced an especially strong Perseid shower on the Mir space station. Robert Zimmerman describes the excitement:

> Placed on 24-hour alert, they watched as about 240 meteors burned up in the atmosphere *below them*. Tsibliev reported "battle wounds" to ground control, while Serebrov described how the solar panels changed color from dark to light blue whenever a micrometeorite hit them. They also spotted 10 new small impacts in the station's windows, including one crater about an inch across. During the height of the storm Mir's sensors indicated a two-thousand-fold increase in the particle flux in the atmosphere surrounding the station. Once, they watched one large meteor burn its way across the northern horizon for more than two seconds.[15]

The cosmonauts' immediate concern was a hit sufficiently big to cause the station to lose its atmosphere. Over time, however, even a mild encounter with micrometeoroids can degrade solar panels, windows, and equipment on the outside of a spaceship.

Another hazard of spaceflight is that human bones tend to lose calcium in zero gravity, although the Russian space crews showed that proper exercise could reduce this tendency. Long-duration human spaceflight might employ centrifugal force to replace gravity, and given that an expedition requires sending a huge amount of food, machinery, and other supplies, this might not be an unreasonable precaution. Two ordinary spacecraft could generate artificial gravity by connecting to each other by long cables, then spinning around their mutual center of gravity, for example. Good health requires wholesome, uncontaminated food, which could present a real challenge on long-duration space missions. NASA Ames has been developing sensors, based on arrays of carbon nanotubes, for monitoring water quality and detecting bacteria.[16] Clearly, sending humans on extended spaceflights will incur many expenses that use of robots will not require.

Several kinds of radiation pose problems not just for humans, but also for machines unless their computer chips have been specially hardened against it. Both Jupiter and Saturn have substantial radiation belts, which would prove lethal for anyone who stayed very long inside them. Once when I was visiting NASA's Jet Propulsion Laboratory, I asked an expert about the radiation environments around the two giant planets. He reported that it would be suicidal to visit Jupiter's moons, and he would not want to go inside Rhea in the systems of Saturn's moons. The optimistic way to read that statement is to suggest that humans should bypass Jupiter altogether and instead focus on Saturn's three largest satellites: Rhea, Titan, and Iapetus. However, solar flares can greatly increase the radiation in interplanetary space, and high-energy cosmic rays are much more common outside the Earth's magnetic field.[17] Quite apart from the short-term risk of radiation poisoning and the long-term risk of cancer, cosmic rays appear to cause irreversible damage to neurons in the brain.[18] Given that all long-duration experience on space stations to date has taken place inside the protection of the Earth's magnetic field, we cannot be sure about the effect on humans of radiation in the interplanetary environment. The report *Nanotechnology in Space Exploration* suggests that nanotechnology might contribute to improved shielding against micrometeoroids and radiation, but getting quickly to the destination is another solution, albeit one that is made complicated by the huge weight of supplies and equipment that must be transported to sustain humans.[19]

THE SOLAR SYSTEM

The future of humans in space depends very much on the actual environment of the solar system, and the extent to which it really can provide economic or psychological rewards. The space program of the past was very much guided by implicit assumptions that may have been wrong. No new assumptions have been discovered that might justify a vigorous space program.

Somewhere across this vast galaxy, there is a solar system with two planets on which life evolved. Intelligence evolved first on one, and then spread to the other via space travel. The first explorers must have had remarkable adventures on the second planet, as they initially battled and ultimately learned to live with the indigenous life forms. Colonists struggled to survive in the alien but biologically rich ecosystem, gradually learning how to tame the wilderness and exploit the abundant natural resources. Trade between the two planets built up a huge infrastructure for interplanetary travel that made it easy to explore and exploit the other more hostile moons and planets of that solar system, and a truly interplanetary civilization was born.

Unfortunately, that solar system is not ours. Venus may be the twin of Earth in size, but its surface conditions are so hot and corrosive that direct exploration by humans is practically impossible, and colonization is quite inconceivable. Mars is also a disappointment, though not entirely beyond the possibility of human habitation. Its atmosphere is so thin that humans would need to wear spacesuits there. The temperature range is comparable to that of the Earth's South Pole, where NSF has shown it is possible to maintain a year-round scientific station, if not to carry out economically profitable activity. But Mars lacks unique natural resources that are not already abundant on the Earth, so it is difficult—albeit not impossible—to imagine a scenario that would warrant its colonization. It is our bad luck that our particular solar system is not ideal for the establishment of interplanetary civilization.

This was not the image that people held of the solar system even 50 years ago. Before it was possible to send space probes to Mars and Venus, relatively little was known about their environments. Many astronomers suspected the truth, but the evidence was weak enough to permit wild speculation. Actually, planetary astronomy was at a low point in its history half a century ago and was revived only when the first space probes were sent. Large traditional telescopes like the ones atop Mount Wilson and Mount Palomar were unable to give clear pictures of the planets, because the Earth's atmosphere blurred the light on its way from space to the telescope mirror. Using short-focus eyepieces to achieve great magnification of the image was not helpful, but merely magnified the atmospheric blur. Thus most astronomers concentrated on the stars and galaxies, using the biggest telescopes to take long-duration time exposures of the most distant objects rather than our closer neighbors in the solar system.

The public image of the planets, to the extent that one existed, was largely shaped by science fiction. The subculture of science fiction writers was born in the late 1920s, mainly reaching its audience through specialty magazines and, to a lesser extent, through novels. Many of the stories they wrote took place in our solar system, during a period imagined to be a few decades or centuries in the future when outposts or colonies had been established on several moons and planets. In addition, authors sometimes wrote about interstellar travel, starting with E. E. "Doc" Smith's "The Skylark of Space" in 1928.[20] Even well into the 1950s, these writers had not yet abandoned the solar system in favor of greener pastures.

Science fiction and popular astronomy in the mass media popularized the notion of planets as being worlds akin to the Earth. Anyone visiting a newsstand would see several monthly science fiction magazines lined up, with their covers depicting humans having adventures on Earth's interplanetary counterparts. Even someone who never read a single one of the stories probably would have seen these magazines, as they remained relatively popular from the late 1920s through the middle of the 1950s, when the fiction magazine industry contracted in the wake of television's growth.[21]

Science fiction movies were rare before *Star Wars* opened the floodgates in 1977, but some did give wide audiences an incorrectly rosy picture of the planets. In particular, four movie serials starring Buster Crabbe, the Olympic swimming star, were extremely popular and widely seen in the period 1936–1940. A family who went to the movies in the era before television would often see a feature film, a newsreel, a couple of cartoons, and one episode of a serial, with typical serials including 12 or 13 episodes. Thus many people who had no interest in spaceflight would have seen at least one episode of *Flash Gordon* in 1936; this serial was set on the planet Mongo, which supposedly had entered the solar system to attack it. *Flash Gordon's Trip to Mars* came in 1938, and in 1940 the hero returned to Mongo in *Flash Gordon Conquers the Universe*. In 1939, the same actor appeared in *Buck Rogers*, which involved a trip to Saturn. Mongo, Mars, and Saturn were all depicted as having humanoid inhabitants and being rich natural environments.

In the United States, television's first demonstration came at the World's Fair in New York in 1939. Although World War II delayed the development of commercial TV, the advent of the atom bomb and other high-technology wartime achievements prepared the public for the remarkable wave of children's science fiction television that emerged over the period 1949–1955. Early adopters of television tended to be prosperous and well educated, so a fair amount of early programming tended to be intellectually pretentious, even as the shows' budgets were low and the television technology itself was still crude. I remember vividly the first television set my family bought in 1948, which featured a slightly brownish black-and-white picture and a screen that was only 10 inches in diagonal measurement. It picked up just four stations from New York City, albeit with a good deal of interference, and my family often encountered difficulty in making all the adjustments with the set's eight knobs. Given these kinds of reception and technology problems that plagued television's early days, science fiction shows could get away with poor special effects so long as the dramas were exciting and contained innovative ideas to entertain the elite audience that television had in the beginning.

During this era, the three dominant children's television programs were *Captain Video* (1949–1955), *Space Patrol* (1950–1955), and *Tom Corbett, Space Cadet* (1950–1955). All took place inside the solar system, although late in their careers all three programs began venturing farther afield.

Captain Video was the first of these shows but least well produced, coming from the ill-fated Dumont network. Around the year 2250, the Captain operated from his hidden mountain retreat, from which he could launch his spaceship, the Galaxy, to deal with problems like the wicked Dr. Pauli. Thus *Captain Video* had parallels with popular radio shows such as *Superman, The*

Shadow, and *The Green Hornet,* in which an elite hero acts largely independently of government direction. Often lacking the budget to put on original dramas, the Captain would tune in to watch the adventures of various Video Rangers, which consisted of episodes from serials of the 1930s and 1940s, including *Flash Gordon* and even some cowboy heroes.

Space Patrol followed the career of Commander "Buzz" Corry of the Space Patrol, played by actor Ed Kemmer (who had been a fighter pilot in World War II), as he protected the United Planets around the year 2950. Based on Robert A. Heinlein's 1948 novel *Space Cadet,* the third of these shows took place around the year 2350 and concerned a team of young men training to be officers in the Solar Guard, which protected the Social Alliance of Earth, Mars, Venus, and other parts of the solar system.

George Pal's series of extremely well-produced science fiction movies showed the general public what real spaceflight might be like. *Destination Moon* (1950) is practically a documentary of the first flight to the Moon, realistically depicting zero gravity in flight, the airlessness of the Moon, and even including a Woody Woodpecker cartoon explaining how a rocket engine generates thrust. *When Worlds Collide* (1951) imagines that a pair of renegade planets is entering the solar system—one destined to destroy the Earth, and the other prepared to provide an alternate home for humanity, if only humans can build spaceships to reach it. In *War of the Worlds* (1953), the Martians come to Earth to seize our planet, but the film begins with a brief lecture about the probable real conditions on the planets of the solar system, including the traditional idea that Mars is a habitable but dying world. The fourth in the series, *Conquest of Space* (1955), suffers from poor writing and acting, but realistically depicts the physics of life on a space station and the first expedition to Mars, where it proves possible to grow plants from Earth. Although these movies suggest that other planets have harsh environments, they imply that humans can master them and that Mars, in particular, would be suitable for colonization.

Any student of spaceflight history knows that a social movement promoted the idea among the American public in the late 1940s and early 1950s. German immigrant Willy Ley, who had participated in the German spaceflight movement, wrote a widely read and accurate history of space technology, which was initially published in 1947 and then updated frequently.[22] Teaming up with artist Chesley Bonestell, he produced *The Conquest of Space* in 1949, which in turn inspired the movie *Destination Moon.*[23] A series of articles in *Collier's,* then a popular weekly magazine, used realistic paintings to popularize the spaceflight ideas of Wernher von Braun, who had run the German V-2 program; the articles were later published as books.[24] Walt Disney also got involved, by commissioning a number of television shows to popularize even further the von Braun model of space exploration. In fact, von Braun and the others in his movement did not claim the planets were especially hospitable, but they did show how spaceflight could be feasible at the current level of scientific development, with the implication that far more could be done after decades and centuries of additional development.

When NASA was founded and a serious space program was built in the United States, this image of the planets—that is, as other worlds that could be visited and even perhaps inhabited—combined with American myths of the frontier and exploration before settlement to strengthen the frankly weak public support for spaceflight. Of course, the idea that the Moon will be another California, and Mars will be another Alaska, is naive in the extreme. Here on Earth, we have even failed to colonize Antarctica despite its huge territory and probable mineral resources. We have the technology to handle the environment of Antarctica, and the cost of travel there is certainly miniscule compared with the cost of a voyage to the Moon or Mars—and yet Antarctica's riches go unexploited. One of the major effects of the converging technologies movement is that raw materials become less valuable, because information is the key. With information technology and nanotechnology, we can create wealth out of the materials immediately at hand. In the 1950s and 1960s, however, people believed that space would be valuable in the same ways that the American west had been a century earlier.[25]

Meanwhile, science fiction seemed to edge away from the original vision of an interplanetary civilization. While the literature remained diverse, some key authors began to move their stories far out into the galaxy. Isaac Asimov's original *Foundation* trilogy concerned a future human civilization spanning the galaxy that had lost track of the planet Earth.[26] Although the original television program *Star Trek* sometimes took place on Earth, it never visited Mars or other locations inside the solar system. Similarly, the *Star Wars* series of movies gave Earth no role whatsoever, and we do not really know if Luke Skywalker and Princess Leia are actually human beings. *Stargate* is built on the premise that an ancient galactic civilization built gates for interstellar travel on many planets; its adventures usually take place many light years from Earth, where the setting is not Earth itself. *Battlestar Galactica*, which first aired in 1978–1980 and was later resurrected in 2003, portrayed Earth as the lost thirteenth colony of humankind that must be sought after Cylon robots have destroyed human civilization on the other known worlds. All these popular mass-media phenomena paint a very ambivalent picture of space exploration, because their planets are often Earthlike and attractive, yet they do not exist in the solar system that we can actually reach.

In his popular book *The Real Mars*, Michael Hanlon shows the public that the red planet is a marvelous place, well worth exploring, but far more hospitable for robots than for humans (Figure 8–1).[27] Amazingly, at the time I write this sentence, the rovers Spirit and Opportunity have nearly completed their third year of life on Mars. An unprotected human, standing on the surface of Mars, would die in five minutes from asphyxiation or in a few seconds if explosive decompression causes massive brain hemorrhages. Working in a spacesuit is difficult, and the suits wear out quickly, so profitable human labor on Mars must probably take place indoors, perhaps in underground cities that robots had already created. Some writers have suggested that it might be possible to terraform Mars—that is, to transform this plant into an Earthlike environment with an ample atmosphere so that humans could live freely on the surface.[28] There is good reason to be skeptical about the economic and technical feasibility of terraforming, but it could not even be attempted until a flourishing economy based on robots had already been established.

Figure 8–1 The view toward the Columbia Hills on Mars, from the robot explorer Spirit (courtesy NASA/JPL/Cornell). This picture was taken in 2004. Later Spirit succeeded in reaching and completing extensive scientific observations on these Martian hills, which were named in memory of the seven astronauts who perished a year earlier when the Columbia space shuttle disintegrated during reentry into the Earth's atmosphere.

The February 2004 report that laid out the new NASA vision for exploration of space gave significant roles to robots, although since then the emphasis seems to have shifted to humans in space rather than machines. To implement this vision, NASA was directed to undertake 19 activities, of which six merely complete the space shuttle assembly of the space station. Six other activities involve robots:

- Undertake lunar exploration activities to enable sustained human and robotic exploration of Mars and more distant destinations in the solar system.

- Starting no later than 2008, initiate a series of robotic missions to the Moon to prepare for and support future human exploration activities.

- Conduct robotic exploration of Mars to search for evidence of life, to understand the history of the solar system, and to prepare for future human exploration.

- Conduct robotic exploration across the solar system for scientific purposes and to support human exploration. In particular, explore Jupiter's moons, asteroids, and other bodies to search for evidence of life, to understand the history of the solar system, and to search for resources.

- Develop and demonstrate power generation, propulsion, life support, and other key capabilities required to support more distant, more capable, and/or longer-duration human and robotic exploration of Mars and other destinations.

- Conduct human expeditions to Mars after acquiring adequate knowledge about the planet using robotic missions and after successfully demonstrating sustained human exploration missions to the Moon.[29]

Perhaps politicians need to think in terms of human astronauts, whose human-interest adventure stories would resonate with voters and might promote greater support for spaceflight programs. I would argue that this is the wrong priority; indeed, we might accomplish so much more over the next half-century if we put our trust in robots rather than politicians. In the human conquest of the universe, robots can be designed to be far more effective than people in the early years. We can travel to the planets both *after* our robots and *as* our robots.

The future of spaceflight may not be the Apollo project redux, but rather may lie in a continuation of the Voyager, Galileo, and Cassini probes to the outer solar system combined with robotic exploration of the Moon and Mars in the heritage of Pathfinder, Spirit, and Opportunity. The human body did not evolve for swimming in the super-cold lakes of Saturn's moon Titan, for digging beneath the icy crust of Jupiter's moon Europa in search of the ocean that may lurk there, or for building the industrial and economic basis for civilization on the arid surface of Mars. Robots can do this work, building cities for humans to inhabit on Mars and the Moon, and acting as our surrogates throughout the solar system.

PERSONALITY TRANSFER

At a 2002 conference sponsored by NASA and held at George Washington University, I proposed that the proper method for human travel to other worlds was not the "spam in a can" model of 1960s astronauts in cramped space capsules, but rather the transfer of human personalities in the form of dynamic information that could come alive by acting through robots, carried within *starbase* information systems, and perhaps by being translated into new biological forms suitable for particular extraterrestrial environments.[30] This would be the fulfillment of converging technologies for improving human performance.

At the first of the Converging Technologies conferences sponsored by NSF, computer graphics pioneer Warren Robinett projected the consequences of understanding how the human brain actually works. He noted:

> If a mind is data that runs on a processor (and its sensors and actuators), then that data—that mind—can travel at the speed of light as bits in a communication path. Thus Mars is less than an hour away at light speed. (We need a rocket to get the first receiver there.) You could go there, have experiences (in a body you reserved), and then bring the experience-data back with you on return.[31]

This may seem a wild idea, beyond any possibility of realistic accomplishment, but in fact a number of scientists and engineers believe that something like it may be entirely possible. For example, the March 2001 issue of *Communications of the Association for Computing Machinery* was devoted to predictions of the next 1,000 years. The very first article concerns *digital immortality*. The authors were Gordon Bell, who is highly respected among computer scientists for his leadership in creating the VAX line of minicomputers, and Jim Gray of Microsoft, who subsequently served on the advisory committee of NSF's Directorate for Computer and Information Science and Engineering. Clearly, digital immortality is not a flaky idea, although we may be rather far from achieving it today. The very first sentence of Bell and Gray's article explains, "Digital immortality, like ordinary immortality, is a continuum from enduring fame at one end to endless experience and learning at the other, stopping just short of endless life."[32]

Back in the 1960s, Roger A. MacGowan and Frederick I. Ordway seriously argued that deep-space exploration and colonization would be carried out by intelligent robots that would become the successor species to humans, as our civilization moved out across the galaxy.[33] More recently, roboticist Hans Morovec of Carnegie Mellon University has argued that intelligent robots would soon become our *mind children*—not merely our successors but entities capable of carrying part or all of our individual personalities.[34] Computer innovator Ray Kurzweil believes this will become possible soon after computers surpass humans in their intellectual abilities, which he projects will occur well before the middle of this century.[35] Notably, in 2003, NASA's chief historian, Steven J. Dick, argued that advanced civilizations across the universe tend to become "postbiological" and based in "artificial intelligence that is a product of *cultural* rather than biological evolution."[36]

The technical ability to transfer with perfect precision a human personality to a robot or information system may not be realized for a century or two, but my research has shown that low-fidelity transfer is possible even today by combining traditional personality measurement methods from psychology with new tools of information management and artificial intelligence from computer science.[37] Thus my vision of the long-term future of technological convergence requires the transformation of human nature, but steps toward that future can be taken today in the manner with which we enter the solar system.

In the context of this chapter, two things are worth noting about this idea. First, any success along these lines will require the full convergence of all NBIC fields, putting nanotechnology, biotechnology, and information technology in the service of new technologies based on cognitive science that remain to be developed. Second, if properly designed, this technology can offer people new justifications for supporting space exploration and establishment of an interplanetary society. This second point especially deserves further consideration.

To archive his or her personality (Figure 8–2), an individual would have to do a good deal of work—not merely coming up with the dollars required to pay professional personality archivists to do their job, but also participating actively in a process of answering thousands of questions and performing in hundreds of tests. Although psychologists have not achieved complete consensus, much scientific literature considers that human personality has five primary dimensions, often called the "Big Five": extraversion, agreeableness, conscientiousness, emotional stability, and intellect.[38] Although 100 questionnaire items are sufficient to determine a person's position along these five

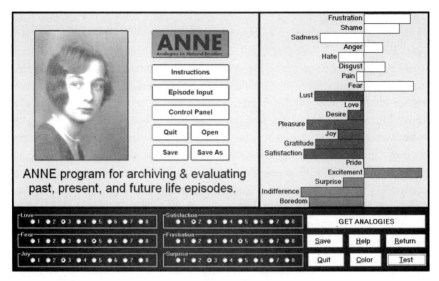

Figure 8–2 One of a series of personality capture modules created by the author. ANNE, or "Analogies in Natural Emotion," is software under development to archive memories of events and to find the similarities that associate some events from a person's life with other events with which they have emotional connections.

dimensions, greater accuracy can be achieved by using 300 to 500 items. Many aspects of character are not measured at all by Big Five items—for example, preferences for foods, activities, and other things that people like or dislike. Thus a comprehensive personality archiving questionnaire should properly include many thousands of items. Having myself responded to more than 100,000 questionnaire items, I would say that it is entirely practical to do so, especially if one can do the work using a pocket or wearable computer at free moments throughout the day. This could easily be done over one or two years of a person's spare time, without precluding leisure activities, but would require great commitment.

For some people, the hope of achieving immortality and traveling into outer space might be motivation enough, but for others some kind of secondary gain would be required. Quite apart from the goal of creating a personality archive, the results of an extensive personality test would provide the individual with insights about his or her fundamental desires, social characteristics, and strategies for action. In this sense, the research subject obtains a secondary gain immediately in terms of greater insight into his or her own character, perhaps learning a few completely new things about himself or herself, but certainly clarifying some previously vague aspects of self-image or identity. This principle should apply to other methods of personality archiving as well. Indeed, another important principle is that a variety of information needs to be collected and preserved, by a corresponding variety of means.

An example of a very different approach to personality transfer might entail use of digital video of the person talking, acting, and emoting. This technique has already become very practical, given that digital video cameras are cheap and many households have the equipment necessary for porting the videos into the computer for annotating, editing, and storage in cheap, long-lasting media. In the future, one collection of videos might depict an individual performing a large variety of ordinary tasks against bland backgrounds and with sharp lighting, so that computer vision techniques could abstract the individual's physical movements. Another excellent approach would be a series of sessions in which the individual describes past experiences of his or her life, perhaps through an interview conducted by an off-camera associate. Some computer vision laboratories and Hollywood special effects studios already have highly accurate, multimedia systems capable of recording speech, gesture, and expression from all sides. In the next few years, we can expect cheap versions to emerge for sale in the home and educational markets.

Before we can outline an agenda that might guide scientific research and technology development, we need to define our goal; in other words, we need to determine precisely what kind of immortality we seek. Many definitions are possible, each of which suggests a particular subset of technological approaches to develop it. The simplest notion is that a person might continue to live in his or her present body indefinitely, using advanced biotechnology assisted by nanotechnology to ward off fatal or degenerative diseases and to halt the natural aging process. Whatever the technical challenges might be in such a case, two moral objections to this goal arise.

First, extending ordinary life indefinitely could be harmful to humanity and to the Earth. The only way to prevent overpopulation would be to radically limit the number of children who could be born. This would, in turn, undercut the very basis of altruism and cooperation, which is nurturance of the young. Survival of the existing population at the expense of unborn lives would be selfish. Reducing the fraction of the population who are children would rob the world of the creativity associated with youth. The solution to this dilemma would be interplanetary and ultimately interstellar migration, such that one might imagine a future society that euthanized any person who remained on Earth after the 100th birthday so as to motivate people to migrate to new worlds. However, humans evolved to fit the environment on Earth, and great effort would be required to prepare suitable homes for us elsewhere so long as we remain in our present form.

This brings us to the second objection: Extending conventional lives could forestall transcendence of the human condition. When religion promises immortality—vainly, I believe—it also promises translation to a higher plane of existence. Western religions believe that the deceased vaults in one leap to the realm of the divine, whereas eastern religions believe in gradual spiritual improvement over a series of incarnations. Even for an atheist, transcendence of current human limitations is an attractive concept, and technologies could be developed to accomplish it. The religious notions of spirit or soul are primitive ways of thinking about aspects of humans that are not simply flesh. Such aspects really exist, and in modern society we call them *information*.

All metaphors distort reality, but they can nevertheless serve as helpful steps toward correct understanding. We can say that human persons are *dynamic systems of information* that happen to be instantiated in flesh but could just as easily be housed in electronics. Human consciousness results from the fact that the individual constantly receives changing information about his or her own actions; computers are capable of similar *feedback*. No one thinks it strange to call parts of a computer its *memory*. Although human brains and contemporary computers use different physical structures to store memory, and different informatic architectures to give meaning to that memory, both are real cases of memory storage. If the human brain is a kind of hardware, then the human personality is a combination of software and memory. Just as a program for one kind of computer can be emulated on another computer, so human software and memory might be ported to different hardware. Although we can imagine reading the contents of a brain directly into a computer, the technology to do so may not be ready for many decades. For now, we need to download the information by analyzing behavior. Answering a questionnaire is one suitable kind of behavior for which the technology is already well developed.

Some of the technology for resurrecting personalities after they have been archived also exists already, although it is not recognized as such, perhaps because it remains fragmentary. For the past two decades, information engineers have been creating expert systems and decision support systems.[39] These databases can be queried. For example, a medical patient can answer questions about his or her symptoms, and the expert system will offer the most likely diagnosis based on the expertise of real doctors that has been stored in it. If you happen to have a file of the personal favorite cooking recipes of a deceased relative, then you have a simple expert system that preserves a little of that person's cooking expertise and embodies some of his or her preferences. In this way, an extensive file of a dead person's preferences (in food, activities, politics, or whatever) can serve as an advisor for living people.

In fact, today's personal computers store considerable information about their users incidentally to performing other tasks—for example, through adaptive help systems, adaptive interfaces, attentive user interfaces, augmented cognition, information filtering, recommender and reputation systems, smart homes, universal access interfaces for the disabled, and user modeling.[40]

Expert systems often employ various kinds of artificial intelligence to make inferences that go beyond the specific pieces of information originally stored, typically by combining existing information in new ways in response to input of new information. Machine learning may also be involved, in which the system adds information that is shaped by the structure and content of the originally stored information. It might be possible to build an expert system based on an individual's detailed political views—and perhaps the fundamental values and personality of the individual—that would be able to extrapolate how that person would respond to new candidates and issues. Long before we can recreate an entire human personality inside a computer system, we will be able to model how that individual would feel and behave in a range of future situations, and even allow that model to take concrete actions such as advising others or voting in elections. Such ability would qualify as a moderate level of immortality, although certainly not full, eternal life.

Another mode of immortality is the creation of a realistic, computerized *avatar*—an animated image of a particular person, in color and three dimensions, with recognizable speech. The motion picture, television, and video game industries have been developing ever more realistic dynamic computer-generated images of human beings, and much fundamental research is investigating the elements required to create realistic avatars, including speech recognition, speech synthesis, language generation, and dialog systems.[41] When combined with other technologies such as expert systems, avatar representation might offer increasingly convincing representations of deceased persons that can then interact with living persons, thereby enriching the lives of the latter and thus becoming part of the living world.

Although we might naively define immortality in terms of indefinite persistence of a human personality as it currently exists, evolution of personalities once they have entered information systems is probably more interesting. A number of radical enhancements are nearly feasible at the present time. If a computerized personality can speak and understand speech in one language, then a translation module would give him or her the ability to do the same in all languages, something that no ordinary human can do.[42] If computer vision gives the archived personality the ability to take in new information visually, then he or she might see images across the entire electromagnetic spectrum, instead of being constrained to the one visual octave (red through blue) available to living humans.[43] Today, by banging on a computer keyboard for a few minutes, people can somewhat arduously find vast sources of new information across the World Wide Web; a computerized personality could have all that information immediately available, as if it were already part of the individual's main memory. More radically, via telepresence a personality could exist simultaneously in several forms—one of which might be exploring the Martian surface in the body of a robot descendant of Spirit and Opportunity, one of which might merge with other archived personalities to become a group mind developing an advanced culture on the far side of the Moon, and one of which might voyage to the stars as the guidance system for a space probe.[44]

WHAT IS TO BE DONE?

To this point, we have considered the innovative work of scientists and engineers, plus the potential for breakthrough research in the future. In truth, everyone has the ability to contribute to this fundamental transformation of human life, especially in the crucial area of info-cogno convergence and the social sciences, where governments are shy about investing research funds. If they can be guided equally by grand visions and by rigorous methodologies, ordinary people can cooperate with professional scientists in preparing the way for a new interplanetary civilization in which transformed humans could live radically advanced lives.

What should a person do? Expertise is always useful, so individuals should read whatever they can find that will inform them about personality-measuring methods. Even an introductory psychology class at a local community college can prove valuable for someone who has not previously studied the subject. Likewise, taking classes or reading textbooks in the research methods of psychology, sociology, political science, and related fields may benefit the ordinary person. One should keep in mind that these fields are still rather "soft" sciences and, therefore, susceptible to pseudo-scientific fads. For example, many companies and government agencies have employed the Myers-Briggs inventory to classify different orientations toward work among employees, although the validity of this instrument is scientifically questionable.[45] An important reason for taking classes and reading textbooks is to acquire a good sense of how complex personality can be, and to realize the need for employing a large number of diverse measures to define any one individual.

Given the low level of government funding for the cognitive and social sciences, ordinary people can contribute their time, effort, and occasionally money to a range of research projects that could provide the knowledgebase for personality preservation or actually demonstrate some of the needed technology. A coalition of students, scientists, and citizens can accomplish much by working collaboratively in this direction, even in the absence of government support. Although an individual can take important but short steps toward immortality through currently available methods for self-archiving, significant progress will require a vigorous social movement.

What should a movement do? A single individual cannot create the necessary technology and systems required for immortality, nor can we rely on such ordinary institutions of society as government agencies, universities, and industrial corporations to handle these tasks. To energize the great cultural shift that is required, a social movement must arise from the interactions of dedicated individuals and evolve into a wholly new societal institution. This movement can serve many functions:

- A framework for the emergence of local, cooperative groups in which individuals help one another archive their personalities
- A communication network that allows widely dispersed individuals to share ideas, information, and inspiration
- An evolving subculture that develops a new set of values concerning the meaning of human life and a fresh paradigm for action
- A source of propaganda that educates the general public and invites talented people to join the cause
- A political activism that overturns laws and regulations that might prevent progress and institutes new ones favoring it
- A leadership structure to coordinate action in accomplishing a variety of needed projects, developing the required infrastructure, and unifying the movement
- A defense against opposition forces, which might include religious and other existing organizations that may feel threatened by change
- A source of political, financial, and labor support for the scientific organizations that will develop the necessary technical basis to transcend the current human condition

Although scientific research is occurring in many of the needed fields, it is not explicitly directed at personality archiving, and the connections between the various projects are weak. Thus one immediate role for a social movement would be to create an inventory of existing projects, perhaps offered in the form of a website that anyone could access, including the researchers themselves. At the same time, members of the movement could conduct their own valuable but low-cost projects doing research in the social, behavioral, and information sciences.

What should scientists and engineers do? With these possibilities in mind, we can sketch a preliminary research and development agenda for scientists and engineers:[46]

- Advanced research at the intersection of the social, cognitive, and information sciences, to improve the reliability, validity, and cost-effectiveness of databases containing results of psychology tests, questionnaires, and the like

- Development of lightweight, nano-enabled, wearable computers that record an individual's natural speech and action while also providing mobile Internet access, information tools, and other benefits that motivate the user to carry one

- Multimedia, multimodal methods for data integration, designed to assemble a variety of information about an individual into a unified model of the person[47]

- Aggressive work in cognitive science, including computational neuroscience, to understand how the human brain actually creates the mind, and thus how to emulate it[48]

- Exploration of new methods, such as nondestructive brain scans, brain–computer interfaces, or nanoscale probes, for harmlessly extracting information about the structure and dynamics of an individual brain[49]

- Improved techniques of anthropometry to measure the shape of a human body, motion capture to preserve its actions, and ways of recording gestures and facial expressions

- Fundamental artificial intelligence research in such areas as neural networks, machine learning, and case-based and rule-based reasoning, to create dynamic models of individuals

- Exploration of how to detect, understand, and emulate human emotions by means of computers and information systems (often called *affective computing*)[50]

- Science and engineering development of animation and display techniques to represent archived personalities to living humans or to each other, through avatars in augmented or virtual reality

- Developments in computer vision, robotics, cyborgs, and related fields to promote perception and action of archived personalities[51]

Although a single person can make a valuable contribution to this movement, he or she cannot do the job alone. The scientific and engineering community can accomplish much, but their energies are currently not harnessed to the task. Therefore, a social movement must arise to weld individuals together, creating a community dedicated to human transcendence of mortality through colonization of the solar system in the form of advanced robots.

CONCLUSION

We should be perfectly honest: Some participants in the converging technologies movement are very skeptical of the advanced possibilities that fascinate the more visionary participants. Some worry that the movement could be stigmatized as crazy, even if its hopes are quite realistic, if it proclaims possibilities that make the average person—or the average politician—nervous. If the aim is to encourage government funding of fundamental research in nanotechnology, for example, it may be wise to keep silent about long-term possibilities while selling policy makers on near-term objectives. At the same time, it would be a tragic mistake to allow the prejudices of unimaginative people to stifle innovation.

Of all the technical controversies around convergence, four stand out:

- Can scientists and engineers really find or create methods for combining historically disparate fields into one universal body of knowledge and capability?
- Can cognitive science really give rise to new technologies and converge with nanotechnology, biotechnology, and information technology?
- Will it be possible to go beyond cognitive science to achieve convergence with transformed social sciences and establish the basis for a unified, peaceful, productive civilization?
- Can convergence ultimately transform the nature of human beings and their position in the cosmos?

Nanoconvergence can provide tremendous benefits for humanity, even if the answers to all four of these questions are negative. Going as far as nature permits us in the direction of convergence will maximize our technical capabilities and sustain economic growth as long as possible—perhaps long enough to bring all human societies to sustainable prosperity. Achieving the vision offered in this chapter—that of a transformed humanity building an interplanetary civilization—requires that all four questions be answered in the affirmative. Even then, some critics of convergence will insist that we must be cautious and stop short of exploiting the possibilities to transform the world and ourselves.

When I contemplate the great questions of convergence and transcendence, two quotations come to mind. One, from Corinthians 13:11 in the New Testament, takes on a new meaning today: "When I was a child, I spake as a child, I understood as a child, I thought as a child: but when I became a man, I put away childish things." It is childish to think that Mars is an Earth-like adventure playground for Flash Gordon, but it is also childish to think that humans will remain close cousins of the chimpanzees as we forge our destiny in the universe. To become worthy inhabitants of the galaxy, we must continue to evolve. The second quotation exists in various forms attributed to the Russian spaceflight pioneer, Konstantin Tsiolkovsky: "The Earth is the cradle of the mind, but we cannot live forever in a cradle." If the Earth is our cradle, then in language borrowed from Luke 2:7, our bodies are our swaddling clothes, and the solar system is our manger. We end with a fifth question, then: "What shall we decide to become?"

REFERENCES

1. Associated Press, "Analysis Adds 'A' to First Words on Moon," *Washington Post*, October 2, 2006, p. A04.

2. William Sims Bainbridge, *The Spaceflight Revolution* (New York: Wiley-Interscience, 1976, p. 3).

3. National Research Council, *An Assessment of Balance in NASA's Science Programs* (Washington, DC: National Academies Press, 2006).

4. William Sims Bainbridge, *The Spaceflight Revolution* (New York: Wiley Interscience, 1976); "Beyond Bureaucratic Policy: The Spaceflight Movement," in James Everett Katz (ed.), *People in Space* (New Brunswick, NJ: Transaction, 1985, pp. 153–163); "The Spaceflight Revolution Revisited," in Stephen J. Garber (ed.), *Looking Backward, Looking Forward* (Washington, DC: National Aeronautics and Space Administration, 2002, pp. 39–64).

5. Mihail C. Roco and William Sims Bainbridge (eds.), *Societal Implications of Nanoscience and Nanotechnology* (Dordrecht, Netherlands: Kluwer, 2001, p. 10).

6. Samuel L. Venneri, "Implications of Nanotechnology for Space Exploration," in Mihail C. Roco and William Sims Bainbridge (eds.), *Societal Implications of Nanoscience and Nanotechnology* (Dordrecht, Netherlands: Kluwer, 2001, p. 214).

7. Nanoscale Science, Engineering, and Technology Subcommittee, Committee on Technology, National Science and Technology Council, *National Nanotechnology Initiative: The Initiative and Its Implementation Plan* (Arlington, VA: National Nanotechnology Coordination Office, 2002, p. 14).

8. Nanoscale Science, Engineering, and Technology Subcommittee, Committee on Technology, National Science and Technology Council, *The National Nanotechnology Initiative: Research and Development Leading to a Revolution in Technology and Industry*, supplement to the President's 2006 budget (Arlington, VA: National Nanotechnology Coordination Office, 2005, p. 11).

9. Nanoscale Science, Engineering, and Technology Subcommittee, Committee on Technology, National Science and Technology Council, *Nanotechnology in Space Exploration* (Arlington, VA: National Nanotechnology Coordination Office, 2006, pp. vii–ix).

10. Nanoscale Science, Engineering, and Technology Subcommittee, Committee on Technology, National Science and Technology Council, *The National Nanotechnology Initiative: Research and Development Leading to a Revolution in Technology and Industry*, supplement to the President's 2007 budget (Arlington, VA: National Nanotechnology Coordination Office, 2006, p. 35).

11. National Research Council, *Evaluation of the National Aerospace Initiative* (Washington, DC: National Academies Press, 2004).

12. John M. Logsdon, "'A Failure of National Leadership': Why No Replacement for the Space Shuttle," in Steven J. Dick and Roger D. Launius (eds.), *Critical Issues in the History of Spaceflight* (Washington, DC: National Aeronautics and Space Administration, 2006, pp. 269–300).

13. Robert Zubrin, *The Case for Mars* (New York: Free Press, 1996).

14. Associated Press, "Space Debris Punched Hole in Shuttle," *CNN.com*, October 6, 2006, online at http://www.cnn.com/2006/TECH/space/10/06/space.shuttle.ap/index.html

15. Robert Zimmerman, *Leaving Earth: Space Stations, Rival Superpowers, and the Quest for Interplanetary Travel* (Washington, DC: Joseph Henry Press, 2003, pp. 339–340).

16. Meyya Meyyappan and Harry Partridge, *Nano and Bio Technology Research at NASA Ames* (Moffett Field, CA: NASA Ames Research Center, 2006).

17. Space Studies Board, *Space Radiation Hazards and the Vision for Space Exploration* (Washington, DC: National Academies Press, 2006).

18. David Shiga, "Future Mars Astronauts Have Radiation on Their Minds," *New Scientist,* September 25, 2006, online at http://space.newscientist.com/article/dn10132-future-mars-astronauts-have-radiation-on-their-minds.html

19. Nanoscale Science, Engineering, and Technology Subcommittee, Committee on Technology, National Science and Technology Council, *Nanotechnology in Space Exploration* (Arlington, VA: National Nanotechnology Coordination Office, 2006, pp. 3–13).

20. Edward Elmer Smith (with Lee Hawkins Garby), "The Skylark of Space," *Amazing Stories,* 3:390–417, August 1928; 3:528–559, September 1928; 3:610–636, 641, October 1928.

21. Ron Goulart, *An Informal History of the Pulp Magazines* (New York: Ace, 1973).

22. Willy Ley, *Rockets and Space Travel* (New York: Viking Press, 1947); *Rockets, Missiles, and Space Travel* (New York: Viking Press, 1951); *Rockets, Missiles, and Men in Space* (New York: Viking Press, 1968).

23. Willy Ley and Chesley Bonestell, *The Conquest of Space* (New York: Viking Press, 1949).

24. Cornelius Ryan (ed.), *Across the Space Frontier* (New York: Viking Press, 1952); Wernher Von Braun, Fred L. Whipple, and Willy Ley, *Conquest of the Moon* (New York: Viking Press, 1953); Willy Ley and Wernher von Braun, *The Exploration of Mars* (New York: Viking Press, 1956).

25. Asif A. Siddiqi, "American Space History: Legacies, Questions, and Opportunities for Future Research," in Steven J. Dick and Roger D. Launius (eds.), *Critical Issues in the History of Spaceflight* (Washington, DC: National Aeronautics and Space Administration, 2006, pp. 433–480).

26. Isaac Asimov, *Foundation* (New York: Gnome Press, 1951); *Foundation and Empire* (New York: Gnome Press, 1952); *Second Foundation* (New York: Gnome Press, 1953).

27. Michael Hanlon, *The Real Mars* (New York: Carroll and Graf, 2004).

28. Michael Allaby and James Lovelock, *The Greening of Mars* (New York: St. Martin's Press, 1984); Kim Stanley Robinson, *Red Mars* (New York: Bantam Books, 1993); Kim Stanley Robinson, *Green Mars* (New York: Bantam Books, 1994); Kim Stanley Robinson, *Blue Mars* (New York: Bantam Books, 1996).

29. National Aeronautics and Space Administration, *The Vision for Space Exploration* (Washington, DC: NASA, 2004).

30. William Sims Bainbridge, "The Spaceflight Revolution Revisited," in Stephen J. Garber (ed.), *Looking Backward, Looking Forward* (Washington, DC: National Aeronautics and Space Administration, 2002, pp. 39–64).

31. Warren Robinett, "The Consequences of Fully Understanding the Brain," in Mihail Roco and William Sims Bainbridge (eds.), *Converging Technologies for Improving Human Performance* (Dordrecht, Netherlands: Kluwer, p. 169).

32. Gordon Bell and Jim Gray, "Futuristic Forecast of Tools and Technologies," *Communications of the ACM*, 44(3):29–30, 2001.

33. Roger A. MacGowan and Frederick I. Ordway III, *Intelligence in the Universe* (Englewood Cliffs, NJ: Prentice-Hall, 1966).

34. Hans Morovec, *Mind Children: The Future of Robot and Human Intelligence* (Cambridge, MA: Harvard University Press, 1988).

35. Ray Kurzweil, *The Age of Spiritual Machines: When Computers Exceed Human Intelligence* (New York: Viking, 1999); *The Singularity Is Near: When Humans Transcend Biology* (New York: Viking, 2005).

36. Steven J. Dick, "Cultural Evolution, the Postbiological Universe and SETI," *International Journal of Astrobiology* 2(11):65, 2003.

37. William Sims Bainbridge, "A Question of Immortality," *Analog*, 122(5):40–49, 2002; "Massive Questionnaires for Personality Capture," *Social Science Computer Review*, 21(3):267–280, 2003; "The Future of the Internet: Cultural and Individual Conceptions," in Philip N. Howard and Steve Jones (eds.), *Society Online: The Internet in Context* (Thousand Oaks, CA: Sage, 2004, pp. 307–324); "Cognitive Technologies," in William Sims Bainbridge and Mihail Roco (eds.), *Managing Nano-Bio-Info-Cogno Innovations: Converging Technologies in Society* (Berlin: Springer, 2006, pp. 207–230).

38. Jerry S. Wiggins (ed.), *The Five-Factor Model of Personality* (New York: Guilford Press, 1996).

39. Jay E. Aronson, "Expert Systems," in William Sims Bainbridge (ed.), *Berkshire Encyclopedia of Human–Computer Interaction* (Great Barrington, MA: Berkshire, 2004, pp. 247–251).

40. See the following essays in William Sims Bainbridge (ed.), *Berkshire Encyclopedia of Human–Computer Interaction* (Great Barrington, MA: Berkshire, 2004): Peter Brusilovsky, "Adaptive Help Systems," pp. 1–3; Alfred Kobsa, "Adaptive Interfaces," pp. 3–7; Ted Selker, "Attentive User Interface," pp. 51–54; David Schmorrow and Amy Kruse, "Augmented Cognition," pp. 54–59; Luz M. Quiroga and Martha E. Crosby, "Information Filtering," pp. 351–355; Cliff Lampe and Paul Resnick, "Recommender and Reputation Systems," pp. 595–598; Diane J. Cook and Michael Youngblood, "Smart Homes," pp. 623–627; Gregg Venderheiden, "Universal Access," pp. 744–750; Richard C. Simpson, "User Modeling," pp. 757–760.

41. See the following essays in William Sims Bainbridge (ed.), *Berkshire Encyclopedia of Human–Computer Interaction* (Great Barrington, MA: Berkshire, 2004): Mary P. Harper and V. Paul Harper, "Speech Recognition," pp. 675–681; Jan P. H. van Santen, "Speech Synthesis," pp. 681–688; Regina Barzilay, "Language Generation," pp. 407–411; Susan W. McRoy, "Dialog Systems," pp. 162–167.

42. Katrin Kirchhoff, "Machine Translation," in William Sims Bainbridge (ed.), *Berkshire Encyclopedia of Human–Computer Interaction* (Great Barrington, MA: Berkshire, 2004, pp. 441–448).

43. Jack M. Loomis, "Sensory Replacement and Sensory Substitution," in Mihail C. Roco and William Sims Bainbridge (eds.), *Converging Technologies for Improving Human Performance* (Dordrecht, Netherlands: Kluwer, 2003, pp. 213–224).

44. William Sims Bainbridge, "The Spaceflight Revolution Revisited," in Stephen J. Garber (ed.), *Looking Backward, Looking Forward* (Washington, DC: National Aeronautics and Space Administration, 2002, pp. 39–64); John V. Draper, "Telepresence," in William Sims Bainbridge (ed.), *Berkshire Encyclopedia of Human–Computer Interaction* (Great Barrington, MA: Berkshire, 2004, pp. 715–719).

45. Daniel Druckman and Robert A. Bjork (eds.), *In the Mind's Eye: Enhancing Human Performance* (Washington, DC: National Academies Press, 1991); David J. Pettinger, "The Utility of the Myers-Briggs Type Indicator," *Review of Educational Research*, 63:467–488, 1993.

46. See the following essays in William Sims Bainbridge (ed.), *Berkshire Encyclopedia of Human–Computer Interaction* (Great Barrington, MA: Berkshire, 2004): James Witte and Roy Pargas, "Online Questionnaires," pp. 520–525; Judith S. Olson, "Psychology and HCI," pp. 586–593; Thad Starner and Bradley Rhodes, "Wearable Computer," pp. 797–802; Dharma P. Agrawal, "Mobile Computing," pp. 452–454; Olufisayo Omojokun Prasun Dewan, "Ubiquitous Computing," pp. 737–741; Rajeev Sharma, Sanshzar Kettebekov, and Gouray Cai, "Multimodal Interfaces," pp. 480–485; Melody M. Moore, Adriane B. Davis, and Brendan Allison, "Brain–Computer Interfaces," pp. 75–80; Victor Paquet and David Feathers, "Anthropometry," pp. 26–32; Jezekiel Ben-Arie, "Motion Capture and Recognition," pp. 457–461; Francis Quek, "Gesture Recognition," pp. 288–292; Irfan Essa, "Facial Expressions," pp. 255–259; Robert St. Amant, "Artificial Intelligence," pp. 40–47; Ira Cohen, Thomas S. Huang, and Lawrence S. Chen, "Affective Computing," pp. 7–10; Jennifer Allanson, "Physiology," pp. 551–554; Abdennour El Rhalibi and Yuanyuan Shen,

"Animation," pp. 13–17; Benjamin C. Lok and Larry F. Hodges, "Virtual Reality," pp. 782–788; Jeremy N. Bailenson and James J. Blascovich, "Avatars," pp. 64–68; Rajeev Sharma and Kuntal Sengupta, "Augmented Reality," pp. 59–64; Erika Rogers, "Human–Robot Interaction," pp. 328–332; William Sims Bainbridge, "Cyborgs," pp. 145–148.

47. Larry Cauller and Andy Penz, "Artificial Brains and Natural Intelligence," in Mihail C. Roco and William Sims Bainbridge (eds.), *Converging Technologies for Improving Human Performance* (Dordrecht, Netherlands: Kluwer, 2003, pp. 256–260).

48. Warren Robinett, "The Consequences of Fully Understanding the Brain," in Mihail C. Roco and William Sims Bainbridge (eds.), *Converging Technologies for Improving Human Performance* (Dordrecht, Netherlands: Kluwer, 2003, pp. 166–170); William A. Wallace, "Engineering the Science of Cognition to Enhance Human Performance," in Mihail C. Roco and William Sims Bainbridge (eds.), *Converging Technologies for Improving Human Performance* (Dordrecht, Netherlands: Kluwer, 2003, pp. 281–294).

49. Rodolfo R. Llinas and Valeri A. Makarov, "Brain–Machine Interface via a Neurovascular Approach," in Mihail C. Roco and William Sims Bainbridge (eds.), *Converging Technologies for Improving Human Performance* (Dordrecht, Netherlands: Kluwer, 2003, pp. 244–251); Miguel A. L. Nicolelis, "Human–Machine Interaction," in Mihail C. Roco and William Sims Bainbridge (eds.), *Converging Technologies for Improving Human Performance* (Dordrecht, Netherlands: Kluwer, 2003, pp. 251–255).

50. Rosalind W. Picard, *Affective Computing* (Cambridge, MA: MIT Press, 1997).

51. Sherry Turkle, "Sociable Technologies," in Mihail C. Roco and William Sims Bainbridge (eds.), *Converging Technologies for Improving Human Performance* (Dordrecht, Netherlands: Kluwer, 2003, pp. 150–158).

Index

BOOKS ONLINE
ENABLED

THIS BOOK IS SAFARI ENABLED

INCLUDES FREE 45-DAY ACCESS TO THE ONLINE EDITION

The Safari® Enabled icon on the cover of your favorite technology book means the book is available through Safari Bookshelf. When you buy this book, you get free access to the online edition for 45 days.

Safari Bookshelf is an electronic reference library that lets you easily search thousands of technical books, find code samples, download chapters, and access technical information whenever and wherever you need it.

TO GAIN 45-DAY SAFARI ENABLED ACCESS TO THIS BOOK:

- Go to **http://www.prenhallprofessional.com/safarienabled**
- Complete the brief registration form
- Enter the coupon code found in the front of this book on the "Copyright" page

If you have difficulty registering on Safari Bookshelf or accessing the online edition, please e-mail customer-service@safaribooksonline.com.

PRENTICE
HALL